Industry 4.0—From Smart Factory to Cognitive Cyberphysical Production System and Cloud Manufacturing

Industry 4.0—From Smart Factory to Cognitive Cyberphysical Production System and Cloud Manufacturing

Editor

Marko Mladineo

MDPI • Basel • Beijing • Wuhan • Barcelona • Belgrade • Manchester • Tokyo • Cluj • Tianjin

Editor
Marko Mladineo
University of Split
Croatia

Editorial Office
MDPI
St. Alban-Anlage 66
4052 Basel, Switzerland

This is a reprint of articles from the Special Issue published online in the open access journal *Energies* (ISSN 1996-1073) (available at: https://www.mdpi.com/journal/energies/special_issues/From_Smart_Factory_to_Cognitive_Cyberphysical_Production_System_and_Cloud_Manufacturing).

For citation purposes, cite each article independently as indicated on the article page online and as indicated below:

LastName, A.A.; LastName, B.B.; LastName, C.C. Article Title. *Journal Name* **Year**, *Volume Number*, Page Range.

ISBN 978-3-0365-5487-7 (Hbk)
ISBN 978-3-0365-5488-4 (PDF)

Cover image courtesy of Marko Mladineo

Contents

About the Editor

Marko Mladineo

Marko Mladineo (Assistant Professor, Ph.D.) holds the positions of Assistant Professor and Chair of Industrial Engineering as part of the Faculty of Electrical Engineering, Mechanical Engineering and Naval Architecture, University of Split, Croatia. His research interests are Industry 4.0/5.0, production systems and production networks, information systems and cyber-physical systems, geographic information systems, smart factories, multiobjective optimization and decision-making, lean management, and business process management. He was involved as a researcher in seven scientific projects funded by the EU, DAAD, Fraunhofer Society, and Croatian Ministry for Science and Education. He is an author/co-author of more than 70 scientific texts: 31 scientific journal papers, 34 scientific conference papers, 2 book chapters, and 3 scientific books. In 2020, he received the Jean-Pierre Brans Award 2020 for his application of the multicriteria decision-making PROMETHEE method. He is 38 years old and married with nine children.

Preface to "Industry 4.0—From Smart Factory to Cognitive Cyberphysical Production System and Cloud Manufacturing"

In the last five years, the new industrial paradigm—Industry 4.0—became a global research trend. Now, it is evolving and moving toward Industry 5.0. From the very beginning, it was clear that the smart factory concept will be its main enabler. The smart factory is seen as a place where virtual and reality meet, supported by the industrial Internet of Things and the cyber-physical production system. Today, a step forward toward the cognitive system has been made. It is something that goes beyond smart, as it is more than big data analytics. The cognitive system can learn and adapt to new situations and to new market demands, completely autonomous or with some human support. These cognitive systems require brand new algorithms and decision-making systems.

This Special Issue addressed some of these problems. The modern multiobjective algorithms and multicriteria decision-making methods are applied to various real-world industrial problems: green and sustainable machining, microscale machining, cyber-physical production networks, the optimization of assembly lines, and cybersecurity. Furthermore, a review of the Industry 4.0 evolution toward Industry 5.0 is presented with a special focus on the people. This increase in the influence of human-centricity became a very important topic, especially in distinguishing Industry 4.0/5.0 from transhumanism ideas.

Marko Mladineo
Editor

Review

From Industry 4.0 towards Industry 5.0: A Review and Analysis of Paradigm Shift for the People, Organization and Technology

Marina Crnjac Zizic [1], Marko Mladineo [1,*], Nikola Gjeldum [1] and Luka Celent [2]

[1] Faculty of Electrical Engineering, Mechanical Engineering and Naval Architecture, University of Split, Rudera Boskovica 32, 21000 Split, Croatia; mcrnjac@fesb.hr (M.C.Z.); ngjeldum@fesb.hr (N.G.)

[2] School of Mechanical and Design Engineering, University of Portsmouth, Portsmouth PO1 3DJ, UK; luka.celent@port.ac.uk

* Correspondence: mmladine@fesb.hr; Tel.: +385-21-305-939

Abstract: The industry is a key driver of economic development. However, changes caused by introduction of modern technologies, and increasing complexity of products and production, directly affect the industrial enterprises and workers. The critics of the Industry 4.0 paradigm emphasized its orientation to new technologies and digitalization in a technocratic way. Therefore, the new industrial paradigm Industry 5.0 appeared very soon and automatically triggered a debate about the role of, and reasons for applying, the new paradigm. Industry 5.0 is complementing the existing Industry 4.0 paradigm with the orientation to the worker who has an important role in the production process, and that role has been emphasized during the COVID-19 pandemic. In this research, there is a brief discussion on main drivers and enablers for introduction of these new paradigms, then a literature-based analysis is carried out to highlight the differences between two paradigms from three important aspects—people, organization, and technology. The conclusion emphasizes the main features and concerns regarding the movement towards Industry 5.0, and the general conclusion is that there is a significant change of the main research aims from sustainability towards human-centricity. At the end, the analysis of maturity models that evaluates enterprises' readiness to introduce features of new paradigms is given as well.

Keywords: Industry 4.0; Industry 5.0; people; Operator 5.0; organization; technology; COVID-19

Citation: Zizic, M.C.; Mladineo, M.; Gjeldum, N.; Celent, L. From Industry 4.0 towards Industry 5.0: A Review and Analysis of Paradigm Shift for the People, Organization and Technology. *Energies* **2022**, *15*, 5221. https://doi.org/10.3390/en15145221

Academic Editor: Wen-Hsien Tsai

Received: 21 June 2022
Accepted: 17 July 2022
Published: 19 July 2022

Publisher's Note: MDPI stays neutral with regard to jurisdictional claims in published maps and institutional affiliations.

1. Introduction

In the last decade, we have been witnesses of transformations inside production systems, especially in the field of digitalization [1]. Information and communication technologies (ICT) are involved in every step of production [2]. This causes the various complexities in several aspects: technological, logistical, organizational, and environmental. A complex transformation process is taking place that needs to be effectively managed. The application of new technologies has significant impact on people at work and in everyday life [3]. People are involved in the process of transforming industrial paradigms, whether they work as workers in the industry or as customers that require a specific product from the industry [4]. As important participants, both workers and customers need to be flexible in adapting to new working conditions and open to learning and sharing knowledge [5]. Another important segment in industry affected by paradigm transformations is organization. Today, the organization is characterized by decentralization, where the decision-making process has been delegated by top managers to workers who are lower-level managers and sometimes to blue-collar workers [6]. The idea behind the decentralization is to quicken the decision-making process by those who have the most information and ability to react in real time. However, to support this idea, the application of new technologies and digitalization is crucial. The data collection and processing bring the proper information [7] so the worker can react fast.

A new industrial paradigm, Industry 5.0, appeared very soon after Industry 4.0 and triggered a debate about the role of, and reasons for, applying the new paradigm. The Industry 4.0 is based on the concept of smart factory, where smart products, machines, storage systems, and data unite in the form of the cyber-physical production systems [8,9]. In the technical aspect, Industry 4.0 has improved the human–machine interaction, but in the socially sustainable aspect, technological transformations of Industry 4.0 should carefully consider the central role of humans [10]. The role and importance of employees was emphasized during the COVID-19 pandemic, and the pandemic itself triggered rethinking of the Industry 4.0 paradigm [11,12]. Consequently, the idea of Industry 5.0 appeared as the extension to Industry 4.0 with social and environmental dimension [13]. On the one side, Industry 5.0 is focused on the workers' skills, knowledge, and abilities to cooperate with machines and robots [14,15], and on the other side, on flexibilities in production processes and environmental impact.

The obstacles to introduce Industry 4.0, arising from technological and organizational points of view, lead to rethinking the process about the shortcomings of the Industry 4.0 approach. Several drivers encourage thinking and discussion about the new paradigm. One of the important drivers is personalized product, where customers participate in product design and production for adaptation of the product to their own needs [16,17]. Other drivers originate from the inability of the SMEs (small and medium enterprises) to implement the Industry 4.0 approach [18,19]. There are lot of studies in the context of Industry 4.0 for SMEs, which are pillars of the economy in many countries because of their contribution to gross domestic product (GDP). The new technologies require high investments and knowledge of how to use them and integrate in existing environments to achieve the best results in production [20]. The important question is: Which technologies should the enterprise use to achieve the best results [21]? This opens the question about the assessment of readiness for changes [22–24]. Whatever question is opened, it is always necessary to start with the key enablers of production. This is where the rethinking process is justified with the aim of highlighting the human-centric perspective.

The topics from this article hold a crucial place in each branch of industry. The purpose of this paper is to discuss the main connections between the basic driving concepts of Industry 4.0 and Industry 5.0, but also to emphasize the key enablers: people, organization, and technology, in theoretical and practical context. The key enablers are intensively studied for the last decade in the context of industry progress. The intention of this work is to achieve better understanding of the appearance of Industry 5.0, referring to the key publications which are cited to highlight the directions of research in this field. As a reminder of past trends, Figure 1 shows the transformations through the paradigms according to the important participants and segments of industries. Additionally, better understanding of the appearance of Industry 5.0 represents the base to rethink real industry processes and possibilities to apply appropriate concepts that best suit specific challenges that industry faces, to achieve the best results in each aspect: people, organization, or technology.

The rest of the paper is organized as follows. Section 2 emphasizes basic driving concepts for each paradigm and the flow of the scientific papers collection. Section 3 introduces the review of existing literature, summarizes main transformations to move towards each paradigm, and represents the connection between key enablers and people, organization, and technology in a practical context. Section 4 discusses the importance of represented connections in Section 3 and readiness to introduce any features of new paradigms. Section 5 gives the concluding remarks.

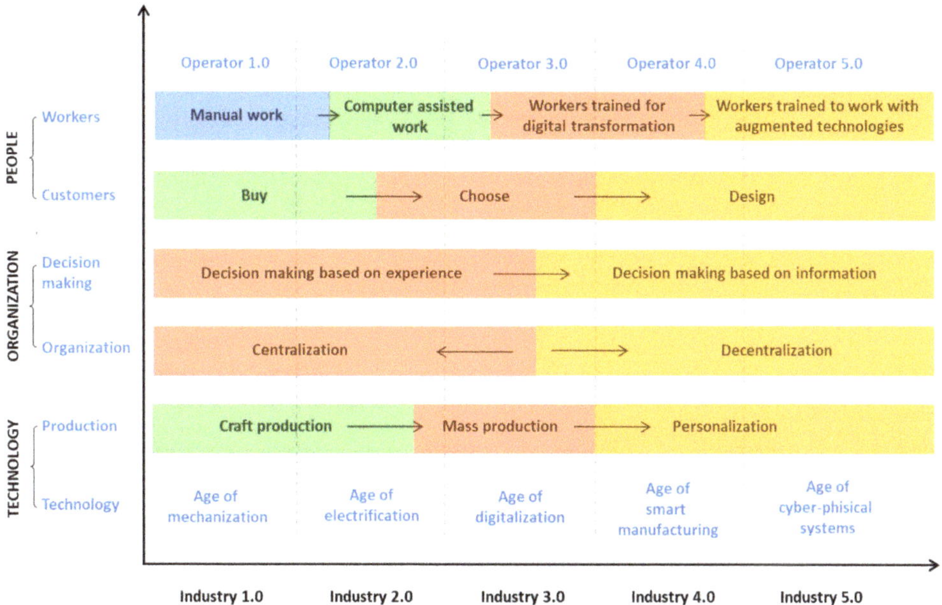

Figure 1. Transformations through the paradigms according to the important participants and segments of industries.

2. Basic Driving Concepts of Industry 4.0 and Industry 5.0

Industry 4.0 is based on the concept of smart factories. The smart factory initiative was founded by partners from industry and academy as an environment for test future technologies [25] and to learn by doing. There are important key drivers [26,27]:

- Internet of Things, services and data that enable the communication between objects. By placing the intelligence into objects, they are turned into smart objects able not only to collect information from the environment and interact or control the physical world, but also to be interconnected to each other through Internet, to exchange data and information [28–30].
- Cloud computing is a driver which supports the Internet of Things, enabling the access to large datasets and its processing to generate new useful information through different types of reports. However, the cybersecurity is a pressing issue; ref. [31] defines cybersecurity as a set of tools, policies and best practices, security concepts, guidelines, risk approaches, actions, assurance, and technologies necessary to protect the cyber environment, organization, and user's assets.
- Cyber-physical system (CPS) is defined as a new generation system with integrated computational and physical capabilities that can interact with people through new modalities [32,33].
- Artificial intelligence supports the cyber-physical system for filtration of the multitude data incoming from different sensors in a production system and analyzes it through the reports. It offers the data-driven predictive analytics and capacity to assist decision-making in highly complex, nonlinear, and multistage production [34,35].
- Augmented reality (AR) represents the integration of the virtual and real environments where objects in the real world are enhanced by computer-generated information or objects with the help of different technologies. AR can be combined with human abilities to provide efficient and complementary tools to assist manufacturing tasks [36].
- Simulation is a powerful tool used for decision making. The application of simulation methods is becoming increasingly relevant as developments in the field of digitaliza-

tion lead to more comprehensive, efficient, embedded, and cost-effective simulation methods [37].

- Autonomous robots can detect problems and independently adjust their tasks to ensure that processes runs smoothly. However, there are levels of robot autonomy, ranging from teleoperation to fully autonomous systems, that influence human–robot interaction [38].

These elements enable the connectivity of the virtual and real world in order to achieve better results in production with maximum profit. A completely profit-driven approach is not sustainable for the long term. Instead of taking technology as a crucial element, the document of European Commission [39] sees three key drivers as the center of new industrial paradigm Industry 5.0 (Figure 2):

- Human-centric approach, which places human needs at the heart of the production process, asking what technology can do for workers and how can it be useful.
- Sustainability, which focuses on reuse, repurpose, and recycle of natural resources and reduce of waste and environmental impact.
- Resilience, which implies an introduction of robustness in industrial production. This robustness provides support through flexible processes and adaptable production capacities, especially when a crisis occurs.

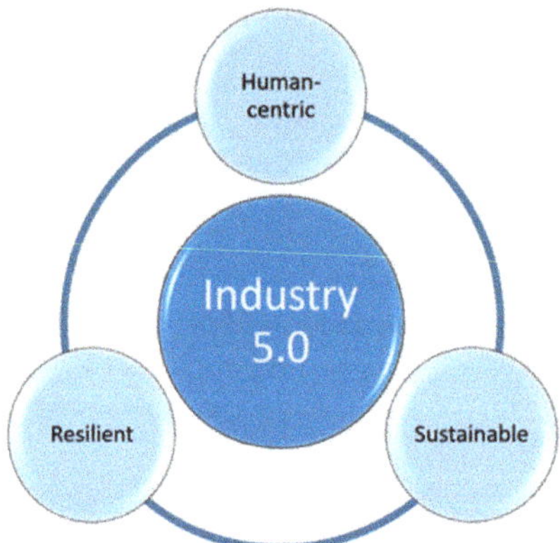

Figure 2. Industry 5.0 with three key drivers.

According to the European Commission, Industry 5.0 is a necessary evolutionary step of Industry 4.0 because of following important issues [40]:

- Industry 4.0 is not the right framework to achieve Europe's 2030 goals, because the current digital economy is a winner-takes-all model that creates technological monopoly and giant wealth inequality.
- Industry 5.0 is not a technological leap forward, but a way to see the Industry 4.0 approach in a broader context, providing regenerative purpose and directionality to the technological transformation of industrial production for people–planet–prosperity.
- Industry 5.0 is a transformative model that reflects the evolution of our thinking post-COVID-19 pandemic, by taking into consideration learnings from the pandemic and the need to design an industrial system that is inherently more resilient to future shocks and truly integrates social and environmental principles.

The next important thing is to identify key enablers in the enterprise, which correlate with the abovementioned drivers of Industry 5.0. Schiele et al. [41] are interpreting Industry 4.0 future within technology, business, society, and people. Similarly, Sony and Naik [42] are investigating the integration of Industry 4.0 with people, infrastructure, technology, processes, culture, and goals. Akcay Kasapoglu [43] is focused on the aspect of leadership and organization during the process of Industry 4.0 transformation. Kayikci et al. [44] are investigating perspectives of people, process, performance, and technology in the Industry 4.0 food supply chain. Kiepas [45] is, similar to Oks et al. [46], simplifying the focus on key enablers, narrowing them to the three most important: humans (people), organization, and technology.

Therefore, a search was carried out within the Scopus database to explore papers related to Industry 4.0/5.0 and three most important enablers: people, organization, and technology (Figure 3). The literature reviews and state-of-the-art papers were excluded from the search.

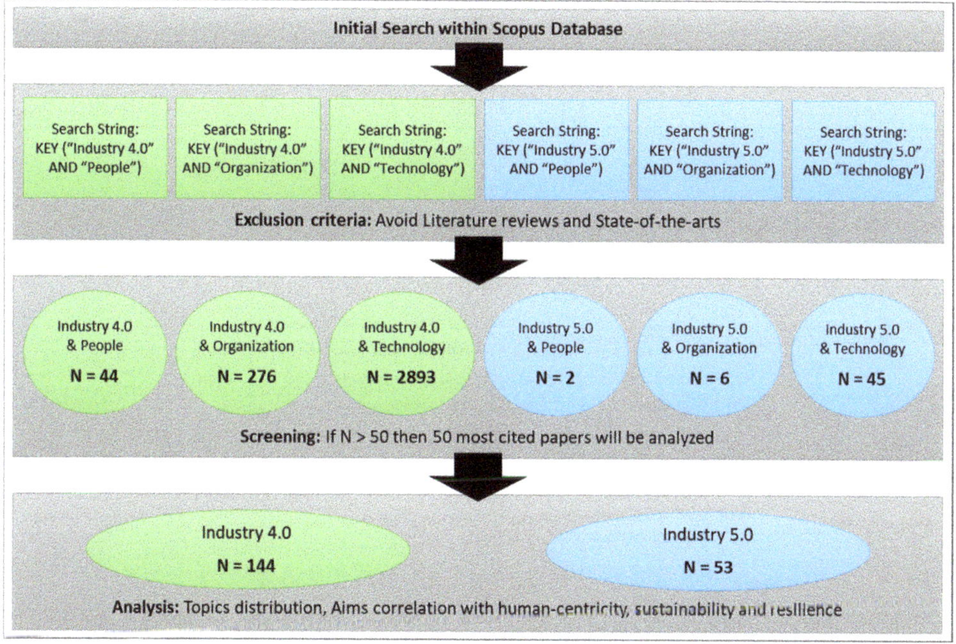

Figure 3. Flow of the scientific papers collection using the Scopus database.

Combinations of Industry 4.0 and Industry 5.0 with each of three enablers resulted in six categories of papers. However, the "Industry 4.0 & Organization" category resulted in 276 papers and "Industry 4.0 & Technology" resulted in 2893 papers, so only the 50 most cited papers were analyzed from each of these two categories. This collection of scientific papers gave scientific perspective to this research.

From the practical real-life perspective, some of the manufacturing industry analyses were used. The analysis of Industry 4.0 implementation in the German manufacturing industry by Veile et al. [47] was inspired by Oks et al. and used the same focus on humans (people), organization, and technology. Furthermore, in the analysis of Croatian manufacturing companies, the questionnaires were given to CEOs to identify basic objectives, main priorities, and the most important aspects regarding how to move towards new industrial paradigms [48]. Again, profound analysis of the results collected with questionnaires identified people, organization, and technology as key enablers [49]. As mentioned, studies have pointed out that each of the three enablers has its important subareas. The most important of them are shown in Figure 4.

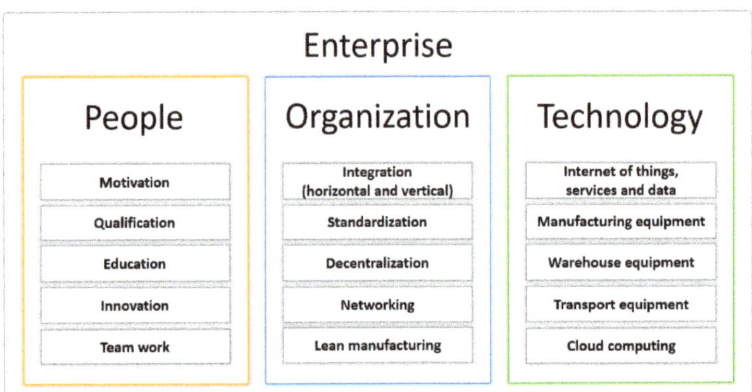

Figure 4. The key enablers to move towards new paradigms: people, organization, and technology.

3. Review of Key Enablers in Practical Context of Industry 4.0 and Industry 5.0

Currently, there are many useful studies about the new technologies characteristic for Industry 4.0, its introduction, and benefits [27]. The studies that emerged from the literature seek a clear vision of how Industry 4.0 impacts business models and organizations [50,51]. However, it is stated how smaller efforts have been devoted to the role of humans in the future factory, the appropriate organizational models, the approaches for long-term value creation, and the outcomes on society [52]. These linked aspects in terms of technology, people, their employability, and sustainability-related issues are crucial for long-term improvement.

Figure 5 summarizes the literature review on Industry 4.0 in relation to people, organization, and technology. The research topics of each enabler were identified and their distribution is presented. Furthermore, the main research aims of analyzed papers were identified in correlation with drivers of Industry 5.0: human-centricity, sustainability, and resilience (Figure 5a).

The same summarization of the literature review on Industry 5.0 is presented in Figure 6. The most interesting fact is a switch in research aims. The sustainability was a major research aim in Industry 4.0, but in Industry 5.0 the human-centricity becomes a major research aim. As already mentioned, the lack of human perspective was a major disadvantage of Industry 4.0 and its main critic [39]. Another interesting fact is the rise of ethical research, with a significant share of research on ethical business and on ethical technology, as well. It is also connected with human-centricity and sustainability of Industry 5.0.

Regarding technology, Industry 4.0 was oriented to every emerging technology. However, with Industry 5.0, some of the emerging technologies are receiving more research focus, and research interest for some others is declining. This change in trends is presented in Figure 7. Again, rise of interest for human–machine interaction and artificial intelligence shows that more focus is being placed on use of technology as a support to the everyday tasks of the human worker. The Industrial Internet of Things has become a major technology research topic, because the 5G cellular network has become a standard nowadays.

3.1. Towards Human-Centricity

In the context of transformation towards Industry 4.0, in the literature exists the awareness that skills, knowledge, and abilities of people are important in the CPS. Operator 4.0 appeared as the operator of the future. The vision of Operator 4.0 aims to create trusting and interaction-based relationships between humans and machines [53]. The ideal type of the factory worker of the future is participative and proactive [54]. There is still a lack of understanding of the interplay between humans and technology [55]. Even though Industry 4.0 was directed to the workers with disabilities (Figure 5), there was not enough adaptation of technology to people. On the one side, people need enough space to develop their skills and use their own creativity, and on the other side, they need technology only as an aid for harmonious collaboration, not to replace their work. This emphasize on human-centricity is now directed towards all workers (Figure 6), not only towards people with disabilities. Operator 5.0 should collaborate with the equipment by using its own physical, sensorial, and cognitive capabilities in an environment that provides safe work and technological assistance in the segments of work that are necessary for the operator, while technologies provide real-time information for making timely decisions.

The limitations for decision-making originate from lack of information, necessary for people for decision-making, which has changed now that the Internet of Things exists. At the moment, there are new technologies and algorithms able to collect huge amounts of data and sort and filter them, in order to use them for decision-making [56–58].

The development of modern tools in ergonomics with the help of Industry 4.0 technologies is becoming of more and more interest in many studies. The International Ergonomics Association offers three main domains of ergonomics: physical ergonomics (working postures, repetitive movements, material handling, safety, and health), cognitive ergonomics (mental workload, decision-making, skilled performances, motor response, trainings, and human–computer interactions), and organizational ergonomics (organizational structures, design of working times, processes, communication, and cooperative work) [59].

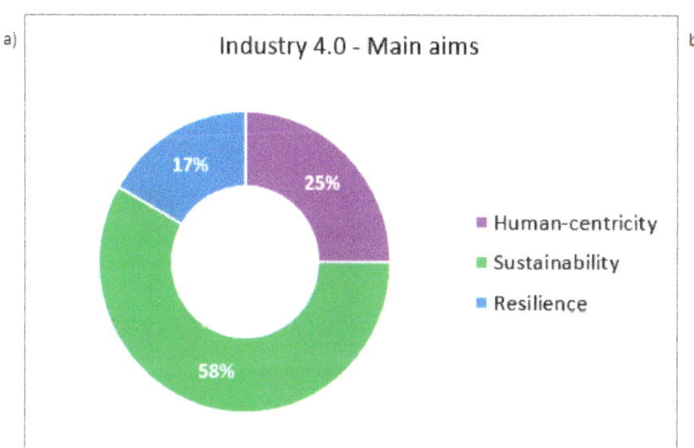

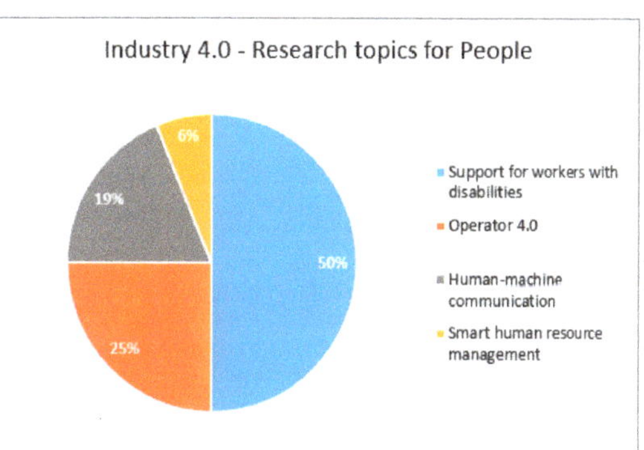

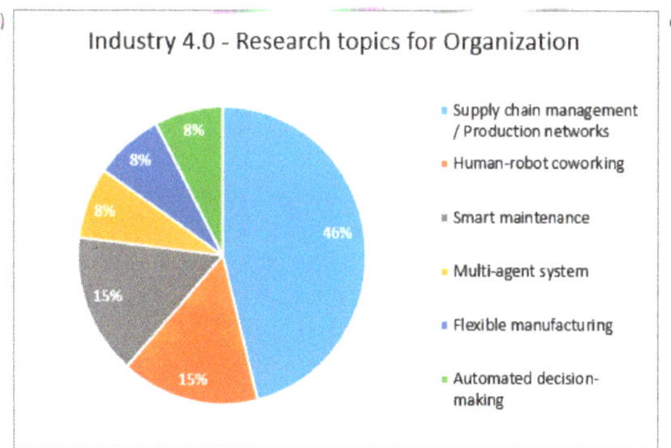

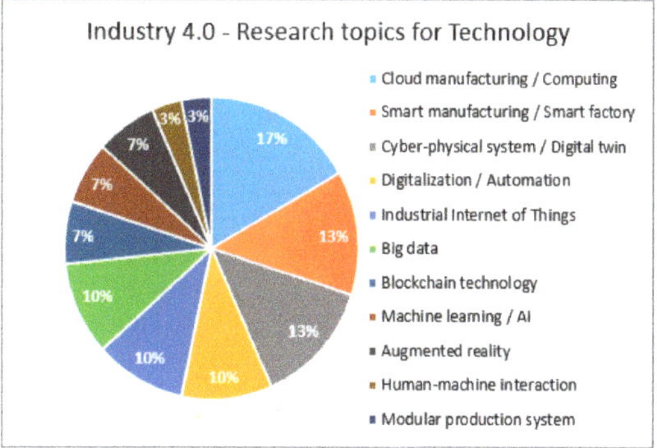

Figure 5. Industry 4.0 literature review: (**a**) Main research aims correlated to Industry 5.0 drivers; (**b**) research topics for Industry 4.0 and people; (**c**) research topics for Industry 4.0 and organization; (**d**) research topics for Industry 4.0 and technology.

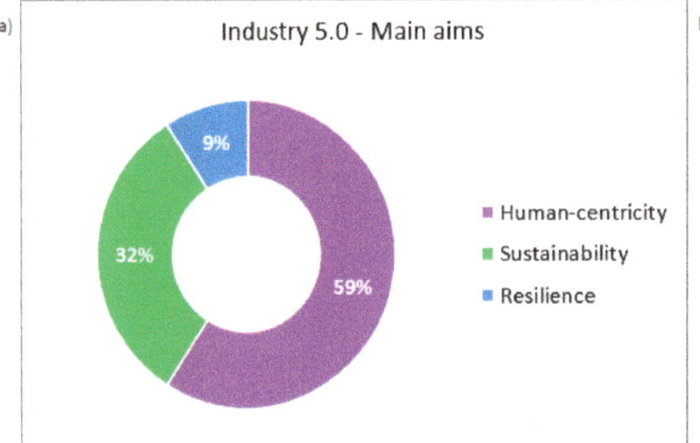

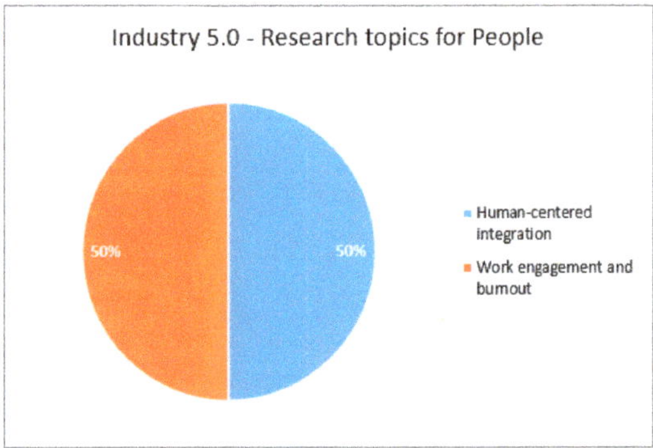

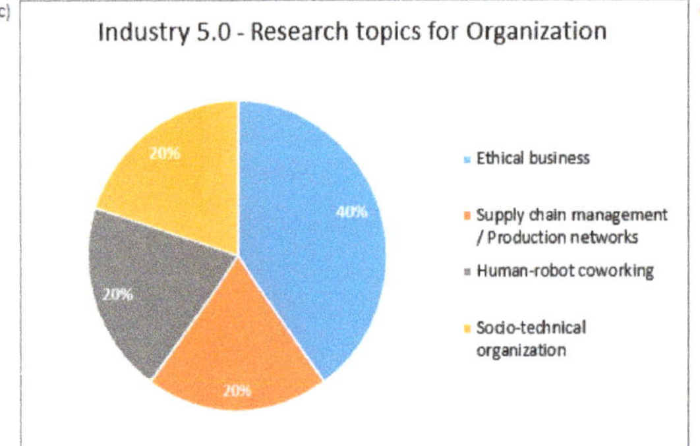

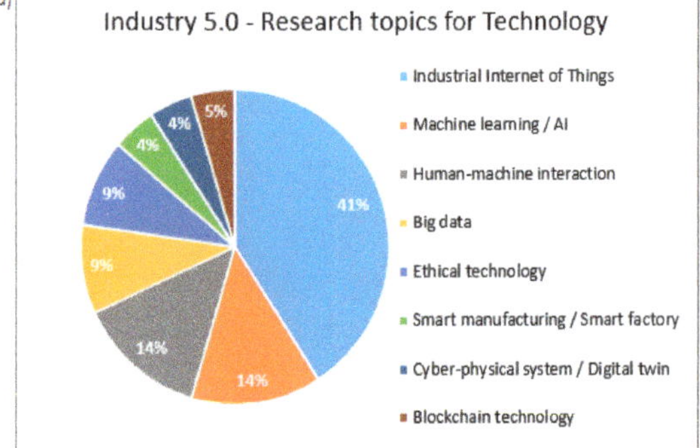

Figure 6. Industry 5.0 literature review: (**a**) Main research aims correlated to Industry 5.0 drivers; (**b**) research topics for Industry 5.0 and people; (**c**) research topics for Industry 5.0 and organization; (**d**) research topics for Industry 5.0 and technology.

The main domains of ergonomics [60]:

- Physical domain—the Industry 4.0 technologies help with the automatization of manual repetitive tasks or hard-muscular tasks [61]; the devices to use on workplace are improving ergonomic feedback and new digital technologies improve internal logistics and transportation [62].
- Cognitive domain—the Industry 4.0 technologies help through virtual models to improve perception and create timely interactions; augmented reality devices contribute to the reduction of mental workload [63]; data sharing is improving cognitive ergonomics.
- Organizational domain—Industry 4.0 provides hybrid production systems to bridge the gap between humans and machines, which affects work organization and requires future skill development.

Digitalization, as a new direction in ergonomics, aims to improve working conditions and the quality of workplaces. The systems to support ergonomics should immediately inform workers about new conditions that appear and about their influence on workers in accordance with analyses that clearly point out requirements for organizationally-technical changes [64]. Other challenges can be found in the studies about the assessment and optimization of postural stress and physical fatigue to identify critical factors and to optimize the assembly operations and workload capabilities at early design stages [65].

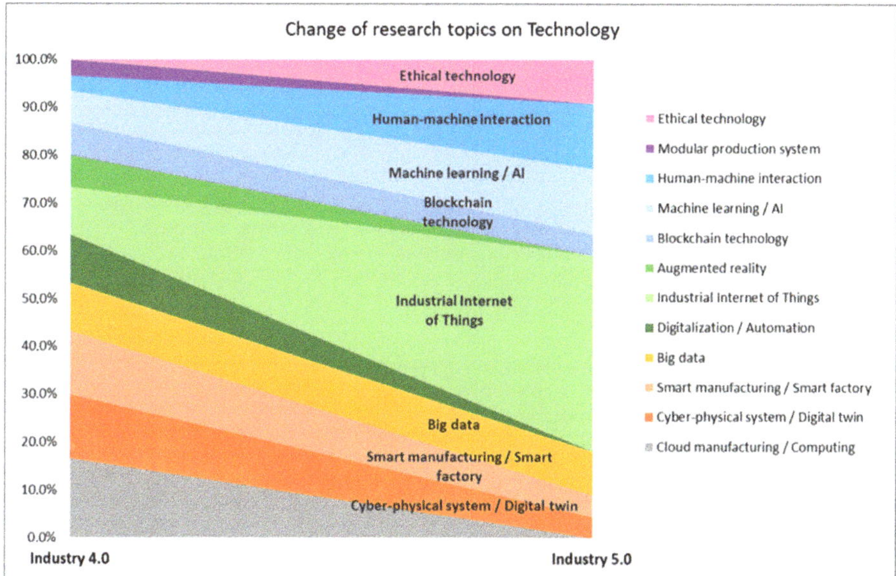

Figure 7. From Industry 4.0 toward Industry 5.0: Change of research topics on technology.

The development of technology, namely, in terms of better adaptation to human needs, is oriented towards the following [66]:

- Networked sensors with low-level intelligence that, at the same time, reduce network overload while allowing exchange of important data.
- Creation of the digital twins, which provides monitoring of production and predicting possible scenarios [67].
- Virtual training for workers to avoid possible dangerous situations while learning specific tasks, for example, in critical review for trainings in construction safety [68], numerous VR/AR systems were proven as efficient, usable, and applicable for training and education; however, there are some challenges to deal with for improvement.
- Artificial intelligence, which enables the learning process for different machines or robots, so they are able to learn from humans and perform tasks based on this knowledge [34,69].

Special attention should be paid to the interaction between humans and robots, machines, or any other elements of the system. There are efforts in the literature to create frameworks for evaluation of the human and robot collaboration. In [70], five dimensions from the aspect of human factor in human–robot collaboration are emphasized for the evaluation: workload, trust, robot morphology, physical ergonomics, and usability. The example of a framework where trajectory prediction serves to avoid potential collisions and plan recognition serves to boost the efficiency of collaboration is included in [71]. The integration of the human–robot collaboration in assembly systems is shown in [72], but taking into consideration operations parameters such as waiting times, parallel activities, and functional delays.

By placing human beings back at the center of industrial production, aided by tools, for example, collaborative robots, Industry 5.0 not only gives consumers the products they want today, but gives workers jobs that are more meaningful than factory jobs have been in well over a century [73]. New jobs are, among other things, aimed at programming, organizing and planning, training, and maintenance. It is clear that knowledge in data science, machine learning, and artificial intelligence is very useful for the future jobs. The fear of jobs disappearing when introducing new technologies is justified because there are situations where automatization of processes with machines or robots can replace human work [74]; however, in the background there are many newly created jobs that enable the introduction of mentioned automatization process [75]. To survive an increasingly competitive market, enterprises need people who are able to manage the changes and should be capable of moving from technology to solutions and from solutions to operations, which requires a broad skillset [76]. Bridging the skills gap requires novel user-facing technologies—such as augmented reality (AR) and wearables—for human performance augmentation to improve efficiency and effectiveness of staff delivered through live guidance [77].

The human-centric approach is firmly attached to Lean management, i.e., its philosophy of the people's engagement in process improvement from shopfloor workers to managers [78]. Lean, as a set of enterprise management tools, represents a strong support for organizing production, managing production, product development, and relationships with suppliers and consumers. To support the sustainable organization, there are studies where Lean management is supported by new technologies characteristic of Industry 4.0 [79]. Lean management is not only based on theory, but is applicable in practice, as many successful companies around the world have proven. Many companies are leading by example, e.g., Toyota production system [80]. They invest very large amounts of money and effort in the development of their own efficient production systems based on lean principles. Many papers carried out research to define frameworks and Lean tools that companies need [81–83]. Design and implementation of such a program of continuous improvement can significantly reduce production losses and the company can be more competitive in the market.

3.2. Towards Sustainability

With adoption of new technologies, it is inevitable to develop new business models. By using smart data, this development has to be exploited for anchoring new, sustainable business models [84]. Designing better business models requires insight into rebound effects and the potential for companies to influence sustainability impacts regarding environmental, social, and economic segments [85]. The main challenges in building the sustainable business model are the balance between profits, social and environmental benefits, reconfiguration of resources and processes for new business models, integration of technologies with business model as a multidimensional and complex task, and usage of the existing business modeling methods and tools [86]. Nevertheless, a significant business model was developed by the European Foundation for Quality Management (EFQM): the EFQM 2020 model [87]. It represents an updated EFQM business excellence model with focus on sustainability, and the model is aligned with the United Nations Sustainable Development Goals. It shares many features with Industry 4.0, especially in the context of transformation and improved organizational performance [88].

In the literature, it is stated how the usage of Industry 4.0 technologies has positive effects on organizational performances. Analysis of the data collected by Duman [89] showed improvements in organizational performance after the usage of Industry 4.0 technologies. Important organizational performance indicators, such as production amount and speed, capacity, quality, and profitability, increased and costs decreased after introduction of the Industry 4.0 technologies. Other studies support this opinion of the positive relationship between Industry 4.0 technologies and organizational performances [90–93].

An important dimension of introduction of technologies related to Industry 4.0 and 5.0 is the efficient usage of energy. New technologies always have an environmental impact, i.e., on the one side, digital technologies demand energy, but on the other side, they save energy. This opens a question: To what extent do industrial paradigms affect environmental sustainability, and is society prepared to deal with those challenges [94]? The adoption of new technologies has negative impacts on the environment, such as air pollution and intensive use of raw materials and energy [95]. However, by adoption of new technologies, the energy can be reduced by analyzing data during the production process and across the supply chain [96]. Furthermore, the process of the technology selection should include environmental and social criteria in order to select technology that is greener and more sustainable, although it can be less productive, at the same time [97].

The important aspect of sustainability and technology is in information logistical waste in the process. The problem with waste of information is recognized during the design and production process and through the supply chain [98,99]. The wastes in the process are related to three parts of data processing: data generation and transfer, data processing, and data storage and data utilization [100]. In data generation, its selection, and evaluation, it is important to collect as much data as necessary and as little as possible to evaluate them according to the content, meaning, and origin. Waste in form of wait periods and data storage matters, particularly regarding data availability in real time. Latencies in the system as well as unprocessed data lead to delays which affect processes. Transfer, movement, and search especially include manual activities, and information is not available in real time, especially when it is written on paper. For data collection or any other work with data, the manual activities should be avoided. The continuous improvement within the manufacturing processes can only be gained by linking and analyzing data.

In existing research about the sustainable energy by using Industry 4.0 technologies, there are data of about 10 to 30 percent of energy reduction for using augmented reality, 5 to 27 percent for using additive manufacturing, around 70 percent of energy savings by using the cloud computing, and 11 to 14 percent global energy reduction using big data and analytics [101].

Another important aspect which has received more attention within Industry 5.0 is the question of ethical use of technology [102,103]. This aspect is closely related to human-centricity, but it is overlapping with sustainability, because low ethical standards produce unsustainable society. Unfortunately, Industry 4.0 also has an ideological aspect in the context of philosophies of transhumanism and posthumanism [104]. These philosophies attack the historical practice in which technology is subordinated to humans, never vice versa. It must be said that, in a way, Industry 4.0 was subordinating human workers to technology. However, the approach of Industry 5.0 is completely ethical and it subordinates technology to the human worker, as the European Commission's document clearly states [39]: "Rather than asking what we can do with new technology, we ask what the technology can do for us".

From Section 3.2. and Figure 5a, it is visible how Industry 4.0 relies on technology to achieve sustainability through different segments of data collection and analysis towards cloud computing. However, from Figure 6a, it is clear how Industry 5.0 is oriented towards human-centricity to resolve mentioned questions of technology adaptation to the human workers, including important ethical aspects.

3.3. Towards Resilience

Resilience, as the ability to withstand disruptions and catastrophic events [105], which relies on people, has not been significantly represented in the concept of Industry 4.0 by the research community. At the moment, there is strong orientation of the literature towards resilience in the context of technology [12], which will be discussed in the below paragraph. It is rarely explored how to rely on people when it comes to resilience. In the developed resilience model by [106], people are one of the most important components because they are the first ones to detect the anomalies and their training and education, awareness building, and leadership, as well as skill and talent, are crucial factors. The strategic human resource management is instrumental in developing requisite knowledge, skills, abilities, and other attributes and in invoking the appropriate collective routines and processes to generate the resilience outcomes [107]. The strategic human resource management can be a critical point [108].

Organizational resilience is a multidisciplinary concept that has its internal and external factors. It represents the ability to overcome the problems caused by internal or external factors. An organizational resilience implies the understanding of the situation, adaption to the new situation, and managing the vulnerabilities. The firm's capacity for developing resilience is derived from a set of specific organizational capabilities, routines, practices, and processes by which a firm orients itself, acts to move forward, and creates a setting of diversity and adjustable integration [107]. Management risk is an important internal factor for resilience. It includes risk plans and prevention techniques. Many studies emphasize information visibility as a crucial factor. Implementation of the key technologies has a positive impact on resilience [109]. New technologies offer the ability to track information that supports organizational resilience. There are positive and negative experiences with introduction of new technology and expectations of it. Each technology should be introduced with special care and consideration about what data it can generate and what benefits it can bring for organization.

Industry 4.0 brought many challenges from the aspects of security, resilience, and efficiency of digital data and systems. Cloud computing is an IT architectural model where computing services are abstracted and delivered to customers on demand, in a self-service way, independent of device and location [110]. Even though there is advanced technology, the information integration across industrial segments, levels, and processes is still a challenge. There are three major integrations in Industry 4.0: horizontal integration, vertical integration, and end-to-end integration. Since the vertical integration represents supply chains, horizontal integration represents collaborative networks [111], also known as production networks, manufacturing networks, and social manufacturing [112]. The COVID-19 pandemic has shown that supply chains can be easily broken, so the collaborative networks are also seen as emergent networks that can replace a broken supply chain and increase the resilience of the manufacturing industry [113].

To achieve integration, it is necessary to change simple information systems to the smart platform [49,112]. In the smart platform there are often high data flow rates and intensive processing requirements, which can cause insufficient system resources for processing to maintain high reliability and resilience [114]. Another challenge is lack of confidence of the industry users in using new technologies, especially from the aspect of data security. To address this, the blockchain technology can support Internet of Things technologies for information exchanges during the different processing stages within a trusted network. Potential applications of blockchain in Industry 4.0 are expected to contribute the following [115]:

- Resilience—being a decentralized peer-to-peer network, blockchain has no single point of failure; it is a durable and immutable ledger; transactions once recorded cannot be altered.
- Scalability—the computing capability of blockchain network scales up as more and more peers join the network.
- Security—all transactions on the blockchain are secured by strong cryptography; as everyone on the network knows about all transactions, they can be easily audited and cannot be disputed.
- Autonomy—blockchain can enable all the components of the CPS to carry out mutual transactions autonomously without the need for a trusted third party; every component has a blockchain account.

However, the blockchain technology is not always an appropriate choice for every firm and has its own challenges, so the need for the blockchain technology in Industry 4.0 can be evaluated according three areas: data exchanges, trusted payments, and data storage [116]. Table 1 summarizes the connection between key enablers for Industry 4.0 and Industry 5.0 and people, organization, and technology.

Table 1. The connection between key enablers for Industry 4.0 and Industry 5.0 and people, organization, and technology in a practical context.

	People	Organization	Technology
Human-centricity	• Operator 5.0 should collaborate with the equipment by using own physical, sensorial, and cognitive capabilities.	• Main domains of ergonomics (physical, cognitive, and organizational). • Digitalization to improve the quality of workplaces.	• Better adaptation of technology to human needs. • Special care regarding the interaction between humans and machines.
• Sustainability	• People who can manage the changes and should be capable of going from technology to solutions and from solutions to operations. • New jobs and knowledge.	• New business models to influence sustainability impacts regarding environmental, social, and economic segment. • Lean management as support.	• Energy reduction. • Efficient usage of energy. • Information logistical waste in the process. • Data analysis for energy reduction.
• Resilience	• People are one of the most important components because they are the first ones to detect the anomalies, and their training and education, awareness building, and leadership, as well as skill and talent are crucial factors.	• A set of specific organizational capabilities, routines, practices, and processes. • Implementation of risk plans and prevention techniques. • Collaborative production networks.	• Decentralized peer-to-peer network. • The information integration across industrial segments, levels, and processes. • Data security. • Smart platforms for collaborative networks.

4. Discussion

The main guideline of why this review and analysis of paradigm shift from Industry 4.0 to Industry 5.0 relies on three segments: people, organization, and technology, comes from existing literature and previous research. In addition, these segments are essential for any manufacturing company, so this paper leads the reader through the basic concepts of each paradigm and essential segments in practical context. The connection between basic concepts of paradigms and these three segments, as is represented in the third section, is important:

- To achieve the goals of each paradigm, where it is crucial to be aware how key enablers are interconnected.
- To review the company's weak points according to the connections of key enablers, so as to know the areas of further action for improvement.
- To rethink the human centricity approach in a company's environment, to adapt technology and organization to people and provide good working conditions as people deserve.
- To assess the sustainability when introducing the change from technological, organizational, or any other aspect.
- To question one's own ability to adapt to changes imposed by either external or internal factors affecting the company.
- To improve a company's organizational performances.
- To strike a balance between effort and investment in change in terms of manpower, organization, and technology.

Another crucial part which follows up on the importance of the connection between the basic concepts of paradigms is the readiness to introduce any feature of new paradigms. The production system or the process that is observed should be ready for that step. A wide range of maturity models are dedicated to the aspects of technical and social systems maturity [117]. The maturity models are increasingly being applied in the area of information technologies [118,119], but especially in digital readiness [120]. There is strong interest from authors in development of maturity models as crucial to adopt new technologies. There are several maturity and readiness models related to Industry 4.0. Through a brief review of existing maturity and readiness models, it is visible that their assessment approaches have a lot in common (Table 2). Models are focused on items for the maturity indication and the range of items' levels. However, in Table 2 it is visible how these models do not cover basic driving concepts characteristic for Industry 5.0. In the models there is a very poor orientation towards human-centricity, sustainability, or resilience.

Table 2. A brief review of some of the existing maturity and readiness models.

Model	Year	Ref.	Approach (Oriented to People (P), Organization (O), and Technology (T), Human-Centric (Hc), Sustainability (S), Resilience (R))	P / Hc	O / S	T / R
The Connected Enterprise Maturity and Readiness Models	2014	[121]	The maturity model is part of the 5 stages (with 5 dimensions) for Industry 4.0. The main focus is on networks, control, working data, analytics, and supply chain relationships.	- / -	O / -	T / -
IMPULS	2015	[122]	The six key dimensions of Industry 4.0 are the foundation for the Readiness model: strategy and organization, smart factory, smart operations, smart products, data-driven services, employees. These six dimensions are used to develop a six-level model for measuring Industry 4.0 readiness.	P / -	O / -	T / -
Digital Operations Self-Assessment	2016	[123]	The model is called "Blueprint for digital success" and it is conducted through 4 stages and 7 dimensions, identifying needs for action as well as classifying current maturity levels. It is focused on digitalization.	P / -	O / -	T / R
Industry 4.0 Maturity Model	2016	[23]	There are nine dimensions in the Industry 4.0 Maturity Model and maturity levels are examined under five levels. Level 1 means that companies have lack of attributes supporting concepts of Industry 4.0, and level 5 means that companies can meet all requirements of Industry 4.0.	P / -	O / -	T / -
SIMMI 4.0 Maturity model	2017	[124]	SIMMI is a System Integration Maturity Model Industry 4.0 which assesses the IT landscape through the four dimensions: vertical integration, horizontal integration, digital product development, cross-sectional technology criteria.	- / -	O / -	T / -
Smart Manufacturing Maturity Model	2018	[125]	There are five dimensions of SME maturity model: finance, people, strategy, process, and product. Technical dimensions are not included in this model. The main focus is on manufacturing operations' performance.	P / -	O / -	- / -
Industry 4.0 technologies assessment	2020	[22]	The maturity model includes the technologies characteristic for Industry 4.0, it allows to compare various technologies in terms of their contribution to the three dimensions of sustainability (economic, environmental, and social).	- / -	- / S	T / -

The assessment of the company's readiness and maturity to introduce any aspect of new paradigms represents the basis. Furthermore, the reasons for introduction are different. In the literature, is stated how the intentions to use Industry 4.0 technologies are more influenced by the expected increase in efficiency than by pressures from suppliers, customers, or even competitors [126]. On the other side, other results yield that the market uncertainty of the business is a significant driver for adoption of Industry 4.0 technologies [127]. For whatever reason, the broader consequences of this introduction should be considered, as well as the human aspect. Additionally, a useful aspect will be the development of a measurement system that can give information about how organization and production benefit from the introduction of the concepts of new paradigms.

5. Conclusions

This review and analysis of paradigm shift for the people, organization, and technology highlights the challenges to introduce the concepts of new paradigms in each essential segment of manufacturing. In contrast with previously published reviews, the goal of this review was to develop a connection matrix between key enablers for Industry 4.0 and Industry 5.0 and essential segments of each manufacturing: people, organization, and technology. People will always be the main drivers of the activities in the production system. The humans that create and manage production systems need support in preparation of infrastructure and resources for introduction of new technologies. In further steps, they need support in transferring the knowledge from a virtual to physical world, and vice versa. This requires future research in domains of adaptation of technology to humans. Significant effort should be made in areas of data collection and interpretation through different useful reports, so that people can make their decisions based on the real-time data. In parallel, there is an effort to create robots that are autonomous and that can collaborate with people. However, these efforts should be made keeping in mind the influence on the sustainability and resilience. On the one side, there are organizational, social, and ergonomic aspects where technology should be at people's disposal, but on the other side, there is energy reduction to satisfy environmental aspects. The crucial aspect is balance between all essential segments in the context of new paradigms, but always keeping human in the center. Generally, the Industry 5.0 paradigm brought the change of main research objectives from sustainability towards human-centricity. From the managerial perspective, it means focusing on workers' education and lifelong learning, instead of focusing on purchase of new technology, or similar. In comparison with Japan and South Korea, the USA and EU are still not investing enough in education of workers, which is becoming an essential issue. Furthermore, it is also important to consider the limitations of this research that are mainly linked to search criteria (keywords). Sometimes the keywords for the paper are not properly selected; thus, applying the keywords filter during the search excludes some high-quality papers, and perhaps includes some papers of lower quality. The future research could use wider search criteria, such as filtering of abstract instead of keywords, in order to create more profound analysis of this topic. Finally, in the future research, the proposed connection matrix can be extended and discussed within the context of specific subareas (i.e., specific subarea for technology can be augmented reality) and all effects of subareas can be seen from the aspect of basic driving concepts and key enablers for each paradigm.

Author Contributions: Conceptualization, M.C.Z. and M.M.; methodology, M.C.Z.; validation, M.C.Z., M.M., N.G. and L.C.; writing—original draft preparation, M.C.Z. and M.M.; writing—review and editing, M.C.Z., M.M., N.G. and L.C. All authors have read and agreed to the published version of the manuscript.

Funding: This research was funded by MZO-VIF INTEMON project.

Institutional Review Board Statement: Not applicable.

Informed Consent Statement: Not applicable.

Data Availability Statement: Not applicable.

Conflicts of Interest: The authors declare no conflict of interest.

References

1. Parida, V.; Sjodin, A.; Reim, W. Reviewing Literature on Digitalization, Business Model Innovation, and Sustainable Industry: Past Achievements and Future Promises. *Sustainability* **2019**, *11*, 391. [CrossRef]
2. Zhou, J. Digitalization and Intelligentization of Manufacturing Industry. *Adv. Manuf.* **2013**, *1*, 1–7. [CrossRef]
3. Ciffolilli, A.; Muscio, A. Industry 4.0: National and Regional Comparative Advantages in Key Enabling Technologies. *Eur. Plan. Stud.* **2018**, *26*, 2323–2343. [CrossRef]
4. Saniuk, S.; Grabowska, S.; Gajdzik, B. Personalization of Products in the Industry 4.0 Concept and Its Impact on Achieving a Higher Level of Sustainable Consumption. *Energies* **2020**, *13*, 5895. [CrossRef]
5. Lepore, D.; Dubbini, S.; Micozzi, A.; Spigarelli, F. Knowledge Sharing Opportunities for Industry 4.0 Firms. *J. Knowl. Econ.* **2021**, *13*, 501–520. [CrossRef]
6. Molino, M.; Cortese, C.G.; Ghislieri, C. The Promotion of Technology Acceptance and Work Engagement in Industry 4.0: From Personal Resources to Information and Training. *Int. J. Environ. Res. Public Health* **2020**, *17*, 2438. [CrossRef]
7. Wang, L.; Wang, G. Big Data in Cyber-Physical Systems, Digital Manufacturing and Industry 4.0. *Int. J. Eng. Manuf.* **2016**, *6*, 1–8.
8. Kagermann, H.; Wahlster, W.; Helbig, J.; Hellinger, A.; Stumpf, M.A.V.; Treugut, L.; Blasco, J.; Galloway, H.; Findeklee, U. *Recommendations for Implementing the Strategic Initiative Industrie 4.0*; National Academy of Science and Engineering: Washington, DC, USA, 2013.
9. Lasi, H.; Fettke, P.; Kemper, H.G.; Feld, T.; Hoffmann, M. Industry 4.0. *Bus. Inf. Syst. Eng.* **2014**, *6*, 239–242. [CrossRef]
10. Kong, X.T.R.; Luo, H.; Huang, G.Q.; Yang, X. Industrial Wearable System: The Human-Centric Empowering Technology in Industry 4.0. *J. Intell. Manuf.* **2019**, *30*, 2853–2869. [CrossRef]
11. Javaid, M.; Haleem, A.; Singh, R.P.; Haq, M.I.U.; Raina, A.; Suman, R. Industry 5.0: Potential Applications in COVID-19. *J. Ind. Integr. Manag.* **2020**, *5*, 507–530. [CrossRef]
12. Xu, X.; Lu, Y.; Vogel-Heuser, B.; Wang, L. Industry 4.0 and Industry 5.0—Inception, Conception and Perception. *J. Manuf. Syst.* **2021**, *61*, 530–535. [CrossRef]
13. Müller, J. *Enabling Technologies for Industry 5.0—Results of a Workshop with Europe's Technology Leaders*; Publications Office of the European Union: Luxembourg, 2020.
14. Demir, K.A.; Döven, G.; Sezen, B. Industry 5.0 and Human-Robot Co-Working. *Procedia Comput. Sci.* **2019**, *158*, 688–695. [CrossRef]
15. Iftikhar, H.M.; Iftikhar, L. Post COVID-19 Industrial Revolution 5.0. The Dawn of Cobot, Chipbot and Curbot. *Pak. J. Surg. Med.* **2020**, *1*, 122–126.
16. Hu, S.J. Evolving Paradigms of Manufacturing: From Mass Production to Mass Customization and Personalization. *Procedia CIRP* **2013**, *7*, 3–8. [CrossRef]
17. Mladineo, M.; Crnjac Zizic, M.; Aljinovic, A.; Gjeldum, N. Towards a Knowledge-Based Cognitive System for Industrial Application: Case of Personalized Products. *J. Ind. Inf. Integr.* **2022**, *27*, 100284. [CrossRef]
18. Horváth, D.; Szabó, R.Z. Driving Forces and Barriers of Industry 4.0: Do Multinational and Small and Medium-Sized Companies Have Equal Opportunities? *Technol. Forecast. Soc. Chang.* **2019**, *146*, 119–132. [CrossRef]
19. Liao, Y.; Deschamps, F.; Loures, E.; de Freitas Rocha Loures, R.; Ramos, L.F.P. Past, Present and Future of Industry 4.0—A Systematic Literature Review and Research Agenda Proposal. *Int. J. Prod. Res.* **2017**, *55*, 3609–3629. [CrossRef]
20. Haseeb, M.; Hussain, H.I.; Ślusarczyk, B.; Jermsittiparsert, K. Industry 4.0: A Solution towards Technology Challenges of Sustainable Business Performance. *Soc. Sci.* **2019**, *8*, 154. [CrossRef]
21. Medić, N.; Anišić, Z.; Lalić, B.; Marjanović, U.; Brezocnik, M. Hybrid Fuzzy Multi-Attribute Decision Making Model for Evaluation of Advanced Digital Technologies in Manufacturing: Industry 4.0 Perspective. *Adv. Prod. Eng. Manag.* **2019**, *14*, 483–493. [CrossRef]
22. Bai, C.; Dallasega, P.; Orzes, G.; Sarkis, J. Industry 4.0 Technologies Assessment: A Sustainability Perspective. *Int. J. Prod. Econ.* **2020**, *229*, 107776. [CrossRef]
23. Schumacher, A.; Erol, S.; Sihn, W. A Maturity Model for Assessing Industry 4.0 Readiness and Maturity of Manufacturing Enterprises. *Procedia CIRP* **2016**, *52*, 161–166. [CrossRef]
24. Rauch, E.; Unterhofer, M.; Rojas, R.A.; Gualtieri, L.; Woschank, M.; Matt, D.T. A Maturity Level-Based Assessment Tool to Enhance the Implementation of Industry 4.0 in Small and Medium-Sized Enterprises. *Sustainability* **2020**, *12*, 3559. [CrossRef]
25. Zuehlke, D. SmartFactory—Towards a Factory-of-Things. *Annu. Rev. Control* **2010**, *34*, 129–138. [CrossRef]
26. Felsberger, A.; Reiner, G. Sustainable Industry 4.0 in Production and Operations Management: A Systematic Literature Review. *Sustainability* **2020**, *12*, 7982. [CrossRef]
27. Oztemel, E.; Gursev, S. Literature Review of Industry 4.0 and Related Technologies. *J. Intell. Manuf.* **2020**, *31*, 127–182. [CrossRef]
28. Borgia, E. The Internet of Things Vision: Key Features, Applications and Open Issues. *Comput. Commun.* **2014**, *54*, 1–31. [CrossRef]
29. Wan, J.; Chin-Feng, L.; Houbing, S.; Muhammad, I.; Dongyao, J. Software-Defined Industrial Internet of Things. *IEEE Sens. J.* **2016**, *16*, 7373–7380. [CrossRef]
30. Wollschlaeger, M.; Thilo, S.; Juergen, J. The Future of Industrial Communication: Automation Networks in the Era of the Internet of Things and Industry 4.0. *IEEE Ind. Electron. Mag.* **2017**, *11*, 7–27. [CrossRef]
31. Tissir, N.; el Kafhali, S.; Aboutabit, N. Cybersecurity Management in Cloud Computing: Semantic Literature Review and Conceptual Framework Proposal. *J. Reliab. Intell. Environ.* **2021**, *7*, 69–84. [CrossRef]
32. Baheti, R.; Gill, H. Cyber-Phisical Systems. *Impact Control Technol.* **2011**, *12*, 161–166.

33. Monostori, L. Cyber-Physical Production Systems: Roots, Expectations and R&D Challenges. *Procedia CIRP* **2014**, *17*, 9–13.
34. Peres, R.S.; Jia, X.; Lee, J.; Sun, K.; Colombo, A.W.; Barata, J. Industrial Artificial Intelligence in Industry 4.0—Systematic Review, Challenges and Outlook. *IEEE Access* **2020**, *8*, 220121–220139. [CrossRef]
35. Patel, P.; Muhammad, I.A.; Amit, S. From Raw Data to Smart Manufacturing: AI and Semantic Web of Things for Industry 4.0. *IEEE Intell. Syst.* **2018**, *33*, 79–86. [CrossRef]
36. Mourtzis, D.; Zogopoulos, V.; Xanthi, F. Augmented Reality Application to Support the Assembly of Highly Customized Products and to Adapt to Production Re-Scheduling. *Int. J. Adv. Manuf. Technol.* **2019**, *105*, 3899–3910. [CrossRef]
37. Sommer, M.; Stjepandic, J.; Stobrawa, S.; von Soden, M. Improvement of Factory Planning by Automated Generation of a Digital Twin. In Proceedings of the Advances in Transdisciplinary Engineering; IOS Press: Amsterdam, The Netherlands, 2020; Volume 12, pp. 453–462.
38. Beer, J.M.; Fisk, A.D.; Rogers, W.A. Toward a Framework for Levels of Robot Autonomy in Human-Robot Interaction. *J. Hum. -Robot. Interact.* **2014**, *3*, 74. [CrossRef]
39. European Commission. *Directorate General for Research and Innovation Industry 5.0—Towards a Sustainable, Human-Centric and Resilient European Industry*; Publications Office of the European Union: Luxemburg, 2021.
40. Renda, A.; Schwaag Serger, S.; Tataj, D. *Industry 5.0, A Transformative Vision for Europe: Governing Systemic Transformations towards a Sustainable Industry*; Publications Office of the European Union: Luxemburg, 2022.
41. Schiele, H.; Bos-Nehles, A.; Delke, V.; Stegmaier, P.; Torn, R.J. Interpreting the Industry 4.0 Future: Technology, Business, Society and People. *J. Bus. Strategy* **2021**, *43*, 157–167. [CrossRef]
42. Sony, M.; Naik, S. Industry 4.0 Integration with Socio-Technical Systems the-Ory: A Systematic Review and Proposed Theoretical Model. *Technol. Soc.* **2020**, *61*, 101248. [CrossRef]
43. Akcay Kasapoglu, O. Leadership and Organization for the Companies in the Process of Industry 4.0 Transformation. *Int. J. Organ. Leadersh.* **2018**, *7*, 300–308. [CrossRef]
44. Kayikci, Y.; Subramanian, N.; Dora, M.; Bhatia, M.S. Food Supply Chain in the Era of Industry 4.0: Blockchain Technology Implementation Opportunities and Impediments from the Perspective of People, Process, Performance, and Technology. *Prod. Plan. Control* **2020**, *33*, 301–321. [CrossRef]
45. Kiepas, A. Humanity–Organization–Technology in View of Industry 4.0. Society 5.0. *Pol. Political Sci. Yearb.* **2021**, *3*, 21–32. [CrossRef]
46. Oks, S.J.; Fritzsche, A.; Möslein, K.M. An Application Map for Industrial Cyber-Physical Systems. In *Industrial Internet of Things: Cybermanufacturing Systems*; Springer: Cham, Switzerland, 2017; pp. 21–46.
47. Veile, J.W.; Kiel, D.; Müller, J.M.; Voigt, K.I. Lessons Learned from Industry 4.0 Implementation in the German Manufacturing Industry. *J. Manuf. Technol. Manag.* **2019**, *31*, 977–997. [CrossRef]
48. Veza, I.; Mladineo, M.; Gjeldum, N. Selection of the Basic Lean Tools for Development of Croatian Model of Innovative Smart Enterprise. *Teh. Vjesn.* **2016**, *23*, 1317–1324.
49. Crnjac, M., Veža, I.; Banduka, N. From Concept to the Introduction of Industry 4.0. *Int. J. Ind. Eng. Manag.* **2017**, *8*, 21–30.
50. Ibarra, D.; Ganzarain, J.; Igartua, J.I. Business Model Innovation through Industry 4.0: A Review. *Procedia Manuf.* **2018**, *22*, 4–10. [CrossRef]
51. Fettig, K.; Gacic, T.; Koskal, A.; Kuhn, A.; Stuber, F. Impact of Industry 4.0 on Organizational Structures. In Proceedings of the 2018 IEEE International Conference on Engineering, Technology and Innovation (ICE/ITMC 2018), Stuttgart, Germany, 17–20 June 2018.
52. Tirabeni, L.; de Bernardi, P.; Forliano, C.; Franco, M. How Can Organisations and Business Models Lead to a More Sustainable Society? A Framework from a Systematic Review of the Industry 4.0. *Sustainability* **2019**, *11*, 6363. [CrossRef]
53. Romero, D.; Bernus, P.; Noran, O.; Stahre, J.; Berglund, Å.F. The Operator 4.0: Human Cyber-Physical Systems & Adaptive Automation towards Human-Automation Symbiosis Work Systems. In Proceedings of the IFIP Advances in Information and Communication Technology; Springer: Cham, Switzerland, 2016; Volume 488, pp. 677–686.
54. Mazali, T. From Industry 4.0 to Society 4.0, There and Back. *AI Soc.* **2018**, *33*, 405–411. [CrossRef]
55. Stentoft, J.; Wickstrøm, K.A.; Philipsen, K.; Haug, A. Drivers and Barriers for Industry 4.0 Readiness and Practice: Empirical Evidence from Small and Medium-Sized Manufacturers. *Prod. Plan. Control* **2021**, *32*, 811–828. [CrossRef]
56. Davidson, R. Cyber-Physical Production Networks, Artificial Intelligence-Based Decision-Making Algorithms, and Big Data-Driven Innovation in Industry 4.0-Based Manufacturing Systems. *Econ. Manag. Financ. Mark.* **2020**, *15*, 16–22.
57. Duft, G.; Durana, P. Artificial Intelligence-Based Decision-Making Algorithms, Automated Production Systems, and Big Data-Driven Innovation in Sustainable Industry 4.0. *Econ. Manag. Financ. Mark.* **2020**, *15*, 9–18.
58. Coatney, K. Milos Poliak Cognitive Decision-Making Algorithms, Internet of Things Smart Devices, and Sustainable Organiza-tional Performance in Industry 4.0-Based Manufacturing Systems. *J. Self-Gov. Manag. Econ.* **2020**, *8*, 9–18.
59. International Ergonomics Association Ergonomic Dimensions. Ergonomic Domains. Available online: https://iea.cc/ (accessed on 12 December 2021).
60. Kadir, B.A.; Broberg, O.; da Conceição, C.S. Current Research and Future Perspectives on Human Factors and Ergonomics in Industry 4.0. *Comput. Ind. Eng.* **2019**, *137*, 106004. [CrossRef]
61. Erol, S.; Jäger, A.; Hold, P.; Ott, K.; Sihn, W. Tangible Industry 4.0: A Scenario-Based Approach to Learning for the Future of Production. *Procedia CIRP* **2016**, *54*, 13–18. [CrossRef]

62. Winkelhaus, S.; Grosse, E.H. Logistics 4.0: A Systematic Review towards a New Logistics System. *Int. J. Prod. Res.* **2020**, *58*, 18–43. [CrossRef]
63. Jeffri, N.F.S.; Awang Rambli, D.R. A Review of Augmented Reality Systems and Their Effects on Mental Workload and Task Performance. *Heliyon* **2021**, *7*, e06277. [CrossRef] [PubMed]
64. Gašová, M.; Gašo, M.; Štefánik, A. Advanced Industrial Tools of Ergonomics Based on Industry 4.0 Concept. *Procedia Eng.* **2017**, *192*, 219–224. [CrossRef]
65. Alkan, B.; Vera, D.; Ahmad, M.; Ahmad, B.; Harrison, R. A Lightweight Approach for Human Factor Assessment in Virtual Assembly Designs: An Evaluation Model for Postural Risk and Metabolic Workload. *Procedia CIRP* **2016**, *44*, 26–31. [CrossRef]
66. Nahavandi, S. Industry 5.0—A Human-Centric Solution. *Sustainability* **2019**, *11*, 4371. [CrossRef]
67. Stjepandić, J.; Sommer, M.; Stobrawa, S. *Digital Twin: Conclusion and Future Perspectives*; Springer: Cham, Switzerland, 2022.
68. Li, X.; Yi, W.; Chi, H.L.; Wang, X.; Chan, A.P.C. A Critical Review of Virtual and Augmented Reality (VR/AR) Applications in Construction Safety. *Autom. Constr.* **2018**, *86*, 150–162. [CrossRef]
69. Kim, D.H.; Kim, T.J.Y.; Wang, X.; Kim, M.; Quan, Y.J.; Oh, J.W.; Min, S.H.; Kim, H.; Bhandari, B.; Yang, I.; et al. Smart Machining Process Using Machine Learning: A Review and Perspective on Machining Industry. *Int. J. Precis. Eng. Manuf. Green Technol.* **2018**, *5*, 555–568. [CrossRef]
70. Gervasi, R.; Mastrogiacomo, L.; Franceschini, F. A Conceptual Framework to Evaluate Human-Robot Collaboration. *Int. J. Adv. Manuf. Technol.* **2020**, *108*, 841–865. [CrossRef]
71. Cheng, Y.; Sun, L.; Liu, C.; Tomizuka, M. Towards Efficient Human-Robot Collaboration with Robust Plan Recognition and Trajectory Prediction. *IEEE Robot. Autom. Lett.* **2020**, *5*, 2602–2609. [CrossRef]
72. Aljinovic, A.; Crnjac, M.; Nikola, G.; Mladineo, M.; Basic, A.; Ivica, V. Integration of the Human-Robot System in the Learning Factory Assembly Process. *Procedia Manuf.* **2020**, *45*, 158–163. [CrossRef]
73. Ostergaard, E. *Welcome to Industry 5.0*; Universal Robots: Odense, Denmark, 2018.
74. Benedikt Frey, C.; Osborne, M.A. *The Future of Employment: How Susceptible Are Jobs to Computerisation?* Oxford Martin Programme on Technology and Employment: Oxford, UK, 2013.
75. Spath, D.; Ganschar, O.; Gerlach, S.; Hämmerle, M.; Krause, T.; Schhlund, S. *Produktionsarbeit Der Zukunft—Industrie 4.0*; Fraunhofer-IAO: Stuttgart, Germany, 2013.
76. Broo, G.D.; Kaynak, O.; Sait, S.M. Rethinking Engineering Education at the Age of Industry 5.0. *J. Ind. Inf. Integr.* **2022**, *25*, 100311. [CrossRef]
77. Ras, E.; Wild, F.; Stahl, C.; Baudet, A. Bridging the Skills Gap of Workers in Industry 4.0 by Human Performance Augmentation Tools—Challenges and Roadmap. In Proceedings of the 10th ACM International Conference Proceeding Series, Rhodes, Greece, 21–23 June 2017; Association for Computing Machinery: New York, NY, USA, 2017; pp. 428–432.
78. Mladineo, M.; Ćubić, M.; Gjeldum, N.; Crnjac Žižić, M. Human-Centric Approach of the Lean Management as an Enabler of Industry 5.0 in SMEs. In Proceedings of the 10th International Conference on Mechanical Technologies and Structural Materials, Split, Croatia, 23–24 September 2021; pp. 111–117.
79. Langlotz, P.; Aurich, J.C. Causal and Temporal Relationships within the Combination of Lean Production Systems and Industry 4.0. *Procedia CIRP* **2020**, *96*, 236–241. [CrossRef]
80. Liker, J. *The Toyota Way*; McGraw-Hill Eductaion: New York, NY, USA, 2003.
81. Belhadi, A.; Touriki, F.E.; el Fezazi, S. A Framework for Effective Implementation of Lean Production in Small and Medium-Sized Enterprises. *J. Ind. Eng. Manag.* **2016**, *9*, 786–810. [CrossRef]
82. Mostafa, S.; Dumrak, J.; Soltan, H. A Framework for Lean Manufacturing Implementation. *Prod. Manuf. Res.* **2013**, *1*, 44–64. [CrossRef]
83. Netland Torbjørn, H. Company-Specific Production Systems: Managing Production Improvement in Global Firms. Ph.D. Thesis, Norwegian University of Science and Technology, Trondheim, Norway, December 2013.
84. Stock, T.; Seliger, G. Opportunities of Sustainable Manufacturing in Industry 4.0. *Procedia CIRP* **2016**, *40*, 536–541. [CrossRef]
85. Bocken, N.; Boons, F.; Baldassarre, B. Sustainable Business Model Experimentation by Understanding Ecologies of Business Models. *J. Clean. Prod.* **2019**, *208*, 1498–1512. [CrossRef]
86. Evans, S.; Vladimirova, D.; Holgado, M.; van Fossen, K.; Yang, M.; Silva, E.A.; Barlow, C.Y. Business Model Innovation for Sustainability: Towards a Unified Perspective for Creation of Sustainable Business Models. *Bus. Strategy Environ.* **2017**, *26*, 597–608. [CrossRef]
87. Fonseca, L. The EFQM 2020 Model. A Theoretical and Critical Review. *Total Qual. Manag. Bus. Excell.* **2022**, *33*, 1011–1038. [CrossRef]
88. Fonseca, L.; Amaral, A.; Oliveira, J. Quality 4.0: The EFQM 2020 Model and Industry 4.0 Relationships and Implications. *Sustainability* **2021**, *13*, 3107. [CrossRef]
89. Calış Duman, M.; Akdemir, B. A Study to Determine the Effects of Industry 4.0 Technology Components on Organizational Performance. *Technol. Forecast. Soc. Chang.* **2021**, *167*, 120615. [CrossRef]
90. Kadir, B.A.; Broberg, O. Human Well-Being and System Performance in the Transition to Industry 4.0. *Int. J. Ind. Ergon.* **2020**, *76*, 102936. [CrossRef]
91. Chauhan, C.; Singh, A.; Luthra, S. Barriers to Industry 4.0 Adoption and Its Performance Implications: An Empirical Investigation of Emerging Economy. *J. Clean. Prod.* **2021**, *285*, 124809. [CrossRef]

92. Kamble, S.S.; Gunasekaran, A.; Ghadge, A.; Raut, R. A Performance Measurement System for Industry 4.0 Enabled Smart Manufacturing System in SMMEs—A Review and Empirical Investigation. *Int. J. Prod. Econ.* **2020**, *229*, 107853. [CrossRef]
93. Büchi, G.; Cugno, M.; Castagnoli, R. Smart Factory Performance and Industry 4.0. *Technol. Forecast. Soc. Chang.* **2020**, *150*, 119790. [CrossRef]
94. Bonilla, S.H.; Silva, H.R.O.; da Silva, M.T.; Gonçalves, R.F.; Sacomano, J.B. Industry 4.0 and Sustainability Implications: A Scenario-Based Analysis of the Impacts and Challenges. *Sustainability* **2018**, *10*, 3740. [CrossRef]
95. Oláh, J.; Aburumman, N.; Popp, J.; Khan, M.A.; Haddad, H.; Kitukutha, N. Impact of Industry 4.0 on Environmental Sustainability. *Sustainability* **2020**, *12*, 4674. [CrossRef]
96. Rocha, A.D.; Freitas, N.; Alemão, D.; Guedes, M.; Martins, R.; Barata, J. Event-Driven Interoperable Manufacturing Ecosystem for Energy Consumption Monitoring. *Energies* **2021**, *14*, 3620. [CrossRef]
97. Celent, L.; Mladineo, M.; Gjeldum, N.; Zizic, M.C. Multi-Criteria Decision Support System for Smart and Sustainable Machining Process. *Energies* **2022**, *15*, 772. [CrossRef]
98. Tribelsky, E.; Sacks, R. Measuring Information Flow in the Detailed Design of Construction Projects. *Res. Eng. Des.* **2010**, *21*, 189–206. [CrossRef]
99. Maurer, C. The Measurement of Information Flow Efficiency in Supply Chain. Ph.D. Thesis, University of South Africa, Pretoria, South Africa, 2013.
100. Meudt, T.; Leipoldt, C.; Metternich, J. Der Neue Blick Auf Verschwendungen Im Kontext von Industrie 4.0 Detaillierte Analyse von Verschwendungen in Informationslogistikprozessen. *ZWF Z. Fuer Wirtsch. Fabr.* **2016**, *111*, 754–758. [CrossRef]
101. Hidayatno, A.; Destyanto, A.R.; Hulu, C.A. Industry 4.0 Technology Implementation Impact to Industrial Sustainable Energy in Indonesia: A Model Conceptualization. *Energy Procedia* **2019**, *156*, 227–233. [CrossRef]
102. Longo, F.; Padovano, A.; Umbrello, S. Value-Oriented and Ethical Technology Engineering in Industry 5.0: A Human-Centric Perspective for the Design of the Factory of the Future. *Appl. Sci.* **2020**, *10*, 4182. [CrossRef]
103. Margherita, E.G.; Braccini, A.M. Managing Industry 4.0 Automation for Fair Ethical Business Development: A Single Case Study. *Technol. Forecast. Soc. Chang.* **2021**, *172*, 121048. [CrossRef]
104. Kravchenko, A. The Forth Industrial Revolution: New Paradigm of Society Development or Posthumanist Manifesto. *Philos. Cosmol.* **2019**, *22*, 128. [CrossRef]
105. Acquaah, M.; Amoako-Gyampah, K.; Jayaram, J. Resilience in Family and Nonfamily Firms: An Examination of the Relationships between Manufacturing Strategy, Competitive Strategy and Firm Performance. *Int. J. Prod. Res.* **2011**, *49*, 5527–5544. [CrossRef]
106. Morisse, M.; Prigge, C. Design of a Business Resilience Model for Industry 4.0 Manufacturers. In Proceedings of the Twenty-Third Americas Conference on Information Systems, Boston, MA, USA, 10–12 August 2017.
107. Lengnick-Hall, C.A.; Beck, T.E.; Lengnick-Hall, M.L. Developing a Capacity for Organizational Resilience through Strategic Human Resource Management. *Hum. Resour. Manag. Rev.* **2011**, *21*, 243–255. [CrossRef]
108. Al-Ayed, S.I. The Impact of Strategic Human Resource Management on Organizational Resilience: An Empirical Study on Hospitals. *Bus. Theory Pract.* **2019**, *20*, 179–186. [CrossRef]
109. Marcucci, G.; Antomarioni, S.; Ciarapica, F.E.; Bevilacqua, M. The Impact of Operations and IT-Related Industry 4.0 Key Technologies on Organizational Resilience. *Prod. Plan. Control* **2021**, 1–15. [CrossRef]
110. Marston, S.; Li, Z.; Bandyopadhyay, S.; Zhang, J.; Ghalsasi, A. Cloud Computing—The Business Perspective. *Decis. Support Syst.* **2011**, *51*, 176–189. [CrossRef]
111. Torn, I.A.R.; Vaneker, T.H.J. Mass Personalization with Industry 4.0 by SMEs: A Concept for Collaborative Networks. *Procedia Manuf.* **2019**, *28*, 135–141. [CrossRef]
112. Mladineo, M.; Celar, S.; Celent, L.; Crnjac, M. Selecting Manufacturing Partners in Push and Pull-Type Smart Collaborative Networks. *Adv. Eng. Inform.* **2018**, *38*, 291–305. [CrossRef]
113. Lagorio, A.; Cimini, C.; Pinto, R.; Paris, V. Emergent Virtual Networks amid Emergency: Insights from a Case Study. *Int. J. Logist. Res. Appl.* **2021**, 1–21. [CrossRef]
114. Xu, L.; Xu, E.L.; Li, L. Industry 4.0: State of the Art and Future Trends. *Int. J. Prod. Res.* **2018**, *56*, 2941–2962. [CrossRef]
115. Joshi, N. Blockchain Meets Industry 4.0—What Happened Next? 2020. Available online: https://www.allerin.com/blog/5659-2 (accessed on 11 November 2021).
116. Fernandez-Caramés, T.M.; Fraga-Lamas, P. A Review on the Application of Blockchain to the Next Generation of Cybersecure Industry 4.0 Smart Factories. *IEEE Access* **2019**, *7*, 45201–45218. [CrossRef]
117. Mettler, T. Maturity Assessment Models: A Design Science Research Approach. *Int. J. Soc. Syst. Sci.* **2011**, *3*, 81–98. [CrossRef]
118. Becker, J.; Knackstedt, R.; Pöppelbuß, J. Developing Maturity Models for IT Management. *Bus. Inf. Syst. Eng.* **2009**, *1*, 213–222. [CrossRef]
119. Becker, J.; Knackstedt, R.; Pöppelbuß, J. Entwicklung von Reifegradmodellen Für Das IT-Management. *Wirtschaftsinformatik* **2009**, *51*, 249–260. [CrossRef]
120. de Carolis, A.; Macchi, M.; Negri, E.; Terzi, S. A Maturity Model for Assessing the Digital Readiness of Manufacturing Companies. In Proceedings of the IFIP International Conference on Advances in Production Management Systems, Hamburg, Germany, 3–7 September 2017; Springer: Cham, Switzerland, 2017; pp. 13–20.
121. Rockwell Automation. *The Connected Enterprise Maturity Model*; Rockwell Automation: Milwaukee, WI, USA, 2014.
122. *IMPULS Industry 4.0 Readiness Online Self-Check for Businesses*; Publications Office of the European Union: Luxemburg, 2017.

123. pwc Industry 4.0—Enabling Digital Operations Self Assessment. Available online: www.pwc.com/industry40 (accessed on 19 September 2021).
124. Leyh, C.; Bley, K.; Schaffer, T.; Bay, L. The Application of the Maturity Model SIMMI 4.0 in Selected Enterprises. In Proceedings of the 23rd Americas Conference on Information Systems (AMCIS 2017), Boston, MA, USA, 10–12 August 2017.
125. Mittal, S.; Romero, D.; Wuest, T. Towards a Smart Manufacturing Maturity Model for SMEs (SM^{3}E). In *IFIP Advances in Information and Communication Technology*; Springer: New York, NY, USA, 2018; Volume 536, pp. 155–163.
126. Čater, T.; Čater, B.; Černe, M.; Koman, M.; Redek, T. Industry 4.0 Technologies Usage: Motives and Enablers. *J. Manuf. Technol. Manag.* **2021**, *32*, 323–345. [CrossRef]
127. Prause, M. Challenges of Industry 4.0 Technology Adoption for SMEs: The Case of Japan. *Sustainability* **2019**, *11*, 5807. [CrossRef]

 *energies*

Article

The Concept of Cyber-Physical Networks of Small and Medium Enterprises under Personalized Manufacturing

Sebastian Saniuk [1,*] and Sandra Grabowska [2,*]

[1]   Department of Engineering Management and Logistic Systems, University of Zielona Gora, 65-417 Zielona Gora, Poland
[2]   Department of Production Engineering, Silesian University of Technology, 40-019 Katowice, Poland
*   Correspondence: s.saniuk@wez.uz.zgora.pl (S.S.); sandra.grabowska@polsl.pl (S.G.)

Abstract: The era of Industry 4.0 is characterized by the use of new telecommunications ICT technologies and networking of the economy. This results in changes both in the way businesses operate and in customer expectations of products offered on the market. The use of modern ICT technologies has made it possible to create cyber-physical systems based on intelligent machines and devices that communicate with each other in real time and allow the integration of resources from different companies to carry out joint production projects. Today's consumer expects products tailored to their needs and expectations. These expectations can be met by leveraging the potential of highly specialized manufacturing service companies centered around e-business platforms. The article presents the results of research using bibliometric analysis and the results of surveys conducted among small and medium-sized enterprises. The concept of e-business platforms supporting rapid prototyping of temporary networks of companies capable of manufacturing personalized products in the environment of Industry 4.0 is presented. The task of the platform is to integrate a customer expecting personalized production with a network of companies having adequate production resources.

Keywords: cyber-physical networks; Industry 4.0; small and medium enterprises; personalization; servitization

Citation: Saniuk, S.; Grabowska, S. The Concept of Cyber-Physical Networks of Small and Medium Enterprises under Personalized Manufacturing. *Energies* **2021**, *14*, 5273. https://doi.org/10.3390/en14175273

Academic Editor: Marko Mladineo

Received: 6 June 2021
Accepted: 24 August 2021
Published: 25 August 2021

Publisher's Note: MDPI stays neutral with regard to jurisdictional claims in published maps and institutional affiliations.

1. Introduction

The Fourth Industrial Revolution introduces new changes in industry, economy and society [1]. Companies operating in today's market are beginning to understand the need for change, especially in the use of modern communication technologies and building a competitive advantage on the market through innovative action. The combination of industrial technologies with modern information and communication technologies (ICT) is the basis of concept proposed by German experts called Industry 4.0 (I 4.0) [2]. This concept is the result of the need to increase the level of industrial production in Western Europe (independence from production coming from the markets of the Middle East) on the one hand, and on the other hand the opportunities offered by the process of digitization and networking of the economy [3]. The Industry 4.0 concept is understood as a joining of intelligent resources and enterprise systems, as well as the introduction of changes in production processes' management which are able to increase production efficiency and flexibility and guarantee a high level of production personalization [4].

Industry 4.0 is called the Fourth Industrial Revolution and is understood as using intelligent technologies in companies and a new approach to people's lives in which mobile devices play an important role in communication. Social networks and unlimited access to information increase consumers' awareness, which results in their growing requirements from products offered on the market [5]. In this case, a modern and innovative approach is needed in the production management of both the enterprise and the network of enterprises that will radically increase flexibility, productivity and customer orientation [6]. According to Reischauer, the Industry 4.0 concept is understood as "policy-driven innovation discourse in manufacturing industries that aims to institutionalize innovation systems that encompass business, academia, and politics. This view clarifies the core identity of Industry 4.0, the intended outcome of Industry 4.0, and the stability of this intended outcome" [7].

Today's customers expect tailor-made products, according to their personal preferences and needs. They expect an impact on the configuration of manufactured products, but at the same time they require a price similar to the products offered in mass production [8,9]. Products are expected to be better tailored to the needs of consumers and even involve them in the design and manufacturing of products through B2C systems. There is a growing demand for personalized products [10].

This means that there is a need for changes in the functioning of modern companies, which must change the business model and focus on the development of service offerings as a complement to the product offerings. In the organization and management sciences, more and more attention is being paid to the importance of services in the production process of goods. Adding services to the modus operandi of modern companies in order to create additional value for the customer is defined as "servitization" [11] or "service infusion" [12,13]. Servitization helps build better customer–manufacturer interaction, makes better use of resources, and provides networking opportunities [14]. This may particularly concern the sector of small and medium enterprises which, unlike large enterprises with great development potential, see an opportunity for development in the conditions of Industry 4.0 in cooperation and narrow specialization [15]. Creating network forms of cooperation is not only an excellent opportunity for manufacturing personalized products and services, but also for dynamizing business models within the concept of Industry 4.0 and a chance to increase the competitiveness of enterprises. The idea of a production network means generating common production orders using fully automated processes of individual network partners, where communication takes place via the internet, and the necessary data is stored in the cloud (cloud computing). The cooperation within the network covers the whole chain of creating value for the customer—from designing the product, through its manufacturing, delivery and use together with offering complementary services (servitization). In the literature, this type of network is also called "Innovative Manufacturing Network of Smart Factories" [16], "Virtual Enterprise Network" [17], "Modular Production Networks" [18] or "Network of Enterprises" [19]. However, focusing on services for networks and development of own know-how requires solving a number of problems resulting from a lack of business models of cyber physical networks of small and medium enterprises [20].

Hence, there is a need for research in the development of business models and concepts for the formation of networks using intelligent resources for the realization of specific personalized products in interaction with the customers. The article emphasizes the need to create e-business platforms that bring together, on the one hand, enterprises offering services and resources for joint implementation of customized production and, on the other hand, customers expecting products tailored to their needs. The main aim of the article is to present the concept of an e-business platform supporting the rapid prototyping of temporary enterprise networks capable of producing personalized products in the Industry 4.0 environment. Furthermore, the article proposes a concept of a methodology for rapid network prototyping that guarantees personalized production orders' execution according to customer specifications. In the paper the following hypothesis: "The organization of e-business platforms of small and medium-sized enterprises will allow the integration of enterprise resources for the formation of cyber-physical networks under the conditions of the fourth industrial revolution" is considered.

2. Challenges of the Fourth Industrial Revolution—State of Research

Industry 4.0 is based on the application of intelligent machines, robots, means of transport and equipment within the cyber-physical system, which contains all activities of product development (product conceptualization, virtual documentation creation, virtual designing, 3D model printing, laboratory and industrial testing, modelling and simulation and lastly production in the real environment) [21]. In addition, the Industry 4.0 concept uses advanced computer-aided systems for design, production management, logistics, sales, service and recycling of products. Dynamic integration of intelligent and autonomous modules of the entire process of production preparation, production and delivery of the product to the customer takes place using IoT technology and information stored in big data and cloud computing. At the same time, a high level of high-efficiency production processes and the meeting of customer expectations is ensured [22]. The Industry 4.0 concept is aimed at significantly improving production efficiency by better using the available resources of enterprises within network cooperation and supply chains [23–25].

Industry 4.0 means the technical integration of all components of cyber-physical systems (CPS) in production and logistics processes through using the Internet of Things (IoT) and the Internet of Services (IoS) [26]. The Industry 4.0 concept requires new business models, reorganization of service and work processes and changes in value chains [27–29].

Under the conditions of the Fourth Industrial Revolution, there is a shift in the manufacturing paradigm towards customized production [30]. The need for customer orientation and offering customers highly personalized products at low prices means the need to change the existing strategies of enterprises. Successively, there is a greater level of interaction between the producer and the customer. Customers are more and more often involved in the creation and even final assembly of a product (e.g., IKEA). In terms of personalization, there are several basic dimensions [31]:

1. One-customer market—personalization is implemented at the individual level so that the customer feels that they are the exclusive or preferred recipient of the service or product.
2. Mass effectiveness—mass customized products are not completely new, as in the case of craft production. In this model, the idea is to customize the product to individual preferences, but with limited fulfillment costs [32].
3. Customer co-creation—companies in today's economy are forced to continually adapt due to increasingly complex and turbulent market conditions. Customer involvement in the design process should not be considered in terms of cost minimization but the chance to obtain higher value. Personalization, in this case, means active interaction with customers. In this case, creating new experiences and building customer satisfaction is more important than just creating a physical product [33].
4. User experience—producers should better understand the hidden needs of the customers instead of the standard exploration of the market potential. Personalization requires that the product be adaptable and configurable in every dimension, i.e., basic structure, design or packaging. It means better meeting the expectations of the individual preferences of the customer [34].
5. Customer-centric companies mean a personalized approach to production. By better integrating the supply chain with the customer, it is possible for the customer to design products online. A highly flexible supply chain will allow for unit production, but on an industrial scale in real-time. This means adopting a production strategy to order, which will lead to very short production lead times and the use of machines with short changeover times [35]. The supply of raw materials and semi-finished products will be based on demand forecasts derived from the demand. Machines will be designed to minimize setup times to accommodate changes in demand, batch sizes, specifications and other parameters [36].

Industry 4.0, by offering incredible technological developments, makes the boundaries between products and services increasingly blurred, enabling a company to transform from a product-based to a service-oriented approach [37]. Servitization represents a significant change in a company's business model, making service activities an engine of growth for the company. The infusion of services takes place when the importance of the services offered increases in relation to the product offer [38].

The concept of servitization usually encompasses enterprises that produce goods to which services are additionally provided. In manufacturing companies, it involves the development and delivery of new services or complex systems that integrate the goods and services provided (product and service solutions). The manufacturing enterprise is moving from exchanging goods with the customer to providing complex solutions [39]. For manufacturing companies, servitization implies a significant change in the perception of the business being run and the vision of its future shape. It is about shifting from a model that focuses on the physical, one-off product sold (possible repairs) to one that relies on regular services around that product [40]. With increasingly complex, high-tech equipment, customers are relying more than ever on their product vendors for expert service. Instead of focusing solely on selling a product, manufacturers are reinventing their strategy to meet the growing needs of customers, resulting in the sale of an entire service system around the product.

Servitization requires significant changes in many areas of a company's operations, and often requires a change in its business model. The product must be seen as a platform for providing a service. It must be accompanied by solutions that customers want. Indeed, these solutions are often captured in product–service systems and product–service combinations [41]. Customers only derive value from them if they actually receive the service—hence, the concept of use value. Servitization can be understood as the process of building revenue streams for producers from services. There are three levels of it that can be offered by producers [42]:

Level I: basic services—goods and spare parts, once they leave the factory, cease to be a problem for the manufacturer; at the same time, they also cease to be a source of revenue.

Level II: intermediate services—product repair, maintenance, overhaul, technical assistance, training, condition monitoring, and product maintenance provide manufacturers with a constant source of revenue.

Level III: advanced services—take after-sales services to a higher level of customer interaction, they are more relationship and customer-oriented than just selling and maintaining the product. In many cases, advanced services are provided in a subscription model in which the customer pays, for example, for hours of music listened to or pages printed.

The potential benefits of competing through advanced services are increased revenues and profits, better alignment with customer needs, enhanced product innovation, building new revenue streams, increased customer loyalty and setting higher barriers to competition [43]. Servitization is usually a subscription model and can be applied to most industries. It is a way to keep companies profitable and competitive in an era where the financial aspects of design and production are increasingly threatened by emerging markets and the life cycle of manufactured products, which is expected to lengthen due to environmental considerations. The need to include additional services, including consulting, is to improve the efficiency and profitability of the company [44]. Producing personalized products and incorporating services into their offerings is designed to meet customer needs [45,46].

3. Materials and Methods

Bibliometric analysis and survey research on a selected group of Polish small and medium-sized enterprises were used to prove the need for research in the field of organization of e-platforms and business models of cooperation of small and medium-sized enterprises within cyber-physical networks.

Bibliometric analysis is a method used to observe the development of science, manifested, inter alia, by creating a network of research connections and the emergence of new, multidisciplinary fields of science. It allows you to identify the internal logic of the development of science [47]. The use of this method in the article made it possible for the reliable assessment of the research results to date and the number of publications and quotations in the analyzed area of research. The bibliometric analysis was conducted within the systematic literature review (SLR) steps [48]. They focus on classification of research contributions and subjective criteria for selecting papers. The search for scientific publications was conducted using the Web of Science (WoS) core collection and Scopus. Based on the methodology adopted in the study, the following steps were proposed: planning, implementation and reporting (Figure 1).

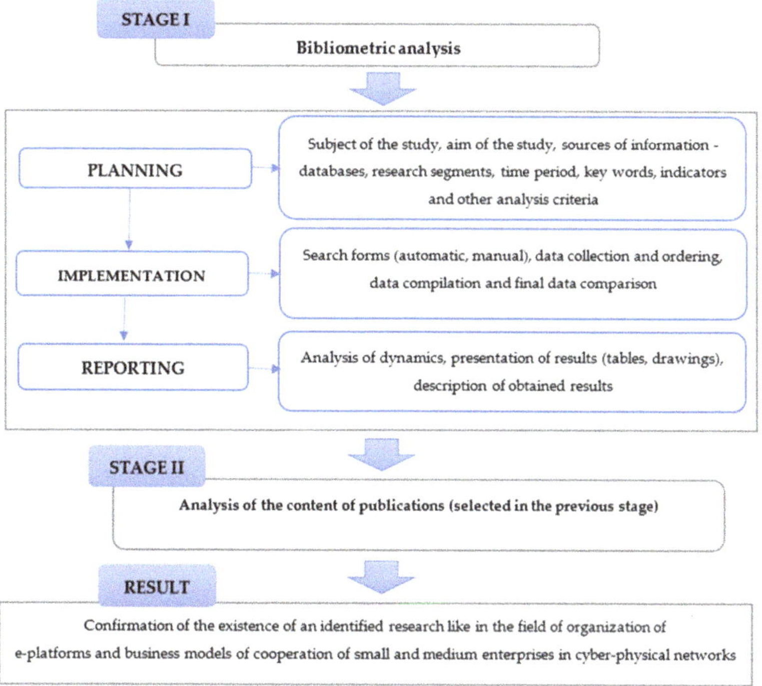

Figure 1. Bibliometric analysis based on methodology SLR. Source: own study.

The search was divided into three stages, each of which led to specific results, thereby increasing knowledge of industrial networks, SME and Industry 4.0. The different stages of the search along with the results they led to are shown in Figure 2.

The bibliometric analysis was intended to provide answers to the following questions:

1. Do authors dealing with industrial networks link this concept to Industry 4.0?
2. Do authors studying the SME sector describe the implementation of Industry 4.0 for this sector?
3. Do the authors dedicate industrial networks to SME?

Furthermore, the research material consisted of the results of a pilot survey conducted among Polish production enterprises from the SME sector. The conducted research concerned small and medium-sized production enterprises representing the metalworking sector. The processes which characterized this sector involved shaping and reshaping metals to create useful objects, parts, assemblies, and different structures. The research was conducted in the period from September 2018 to January 2019. The research was carried out using the CAWI method (computer-assisted web interview), and 50 enterprises took part in the survey (the sample was purposive). All small and medium enterprises agreed to fill the questionnaire and participated in the survey. The only criterion for selecting companies for the study was their size and sector of industry (production). The questionnaire was validated. The pilot surveys were conducted among 15 experts of production management. The questionnaire was revised with their comments. The questionnaire consisted of 12 questions. For the presentation of the results, several answers were selected that identified expectations and barriers related to the introduction of the Industry 4.0 concept and future cooperation of small and medium-sized enterprises. The main purpose of the survey was to demonstrate the need to develop the concept of rapid prototyping of cyber-physical production networks and the terms of cooperation within the network proposed in the article.

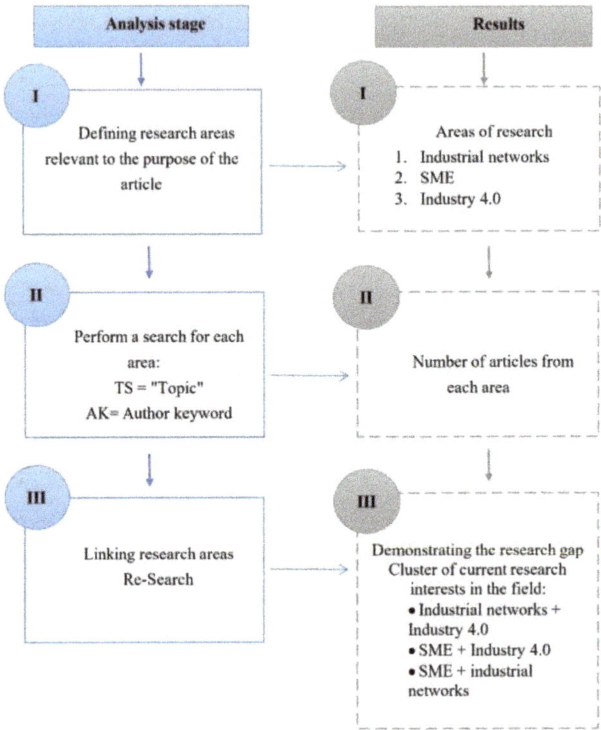

Figure 2. Search stages with results. Source: own elaboration.

4. Results

4.1. Bibliometric Analysis—Results

The body of knowledge related to SME research, industrial networks and Industry 4.0 is spreading significantly. The bibliometric analysis focused precisely on these three thematic/research areas. The focus was exclusively on academic articles, published between 1990 and 2020, in English. The beginning of the search period was defined as 1990, as this was the beginning of the Third Industrial Revolution, in which automation played a major role, enabling industrial networks.

For each research area, articles were searched under the category "Topic" (TS), which includes title, abstract, keywords defined by the authors; and "KeyWords Plus" (words and phrases extracted from the titles of cited articles). This allowed us to show the number of publications from each area. In the further part of the analysis, the search was narrowed down to articles in which the authors indicated as keywords (AK): industrial networks, SME and Industry 4.0, which guarantees the concentration of the article around this specific topic. The research areas were then combined to see how many articles were published in relation to "industrial networks AND Indsutry 4.0", "SME AND Industry 4.0", and "SME AND industrial networks". Detailed search data are included in Table 1.

Table 1. Detailed search data from Web of Science and Scopus databases.

	Database	
	Web of Science	**Scopus**
Year range	1990–2020	1990–2020
Languages	English	English
Type of document	Article	Article
Field Tags:	Number of articles	
	"Industrial networks"	
TS = "industrial networks"	432	836
AK = "industrial networks"	114	387
	"SME"	
TS = "SME"	8.995	22.111
AK = "SME"	1.990	8.647
	"Industry 4.0"	
TS = Industry 4.0	5.371	3.877
AK = Industry 4.0	1.220	2.615
	Advanced Search	
TS = "industrial networks" AND "Industry 4.0"	22	30
AK = "industrial networks" AND "Industry 4.0"	2	7
TS = "SME" AND "Industry 4.0"	44	138
AK = "SME" AND "Industry 4.0"	11	55
TS = "SME" AND "industrial networks"	5	16
AK = "SME" AND "industrial networks"	0	2

Of the three research areas searched, the largest group were articles relating to small and medium enterprises (WoS—TS:8.995, AK: 1.990; Scopus—TS: 22.111, AK:8.647). A smaller group were articles on Industry 4.0 (WoS—TS:5.371, AK; 1.220; Scopus—3.877; AK: 2.615). The smallest group was the articles from the field of "industrial networks" (WoS—TS: 432, AK: 114; Scopus—TS: 836, AK: 387).

Through the bibliometric analysis, it was found that industrial networks are combined with Industry 4.0, which is due to the fact that Industry 4.0 technologies are dedicated to this type of collaboration. Although the research area of SMEs is the largest, the issues of industrial networks and Industry 4.0 are rarely addressed in relation to SMEs; this indicates a definite research gap.

The final stage of the bibliometric analysis was a qualitative analysis of the articles that were extracted during the advanced search. It provided information—what issues are addressed by their authors. Thus, for the search:

- AK = "industrial networks" AND "Industry 4.0" (2 and 7 articles)—the following issues were identified: Data management, teaching and learning, internet of things, Blockchain, Cybersecurity, Digital twin, Digitalization, Supply chain, Command and control, covert channel, data exfiltration, stealth attacks, Distributed Systems, Internet of Things, manufacturing, data management, Performance analysis.
- AK = "SME" AND "Industry 4.0" (11 and 55 articles)—the following issues were identified: production control, smart manufacturing, innovation, digitalization, automation, internet of things, simulation and modeling, e-Services, Supply Chains, Business Model, Mass customization, innovation, Internet of things, Cyber physical systems.
- AK = "SME" AND "industrial networks" (0 and 2 articles)—the following problems were identified: cluster, collaboration, Productivity.

4.2. Results of a Survey of Small and Medium-Sized Enterprises

On the basis of the survey, key problems and expectations of selected companies were identified, which include the implementation of the Industry 4.0 concept in their strategy. On the one hand, the surveyed companies plan to adapt to Industry 4.0 technology in the future, and on the other hand, they do not have a strictly developed implementation plan (92% of respondents). Interesting from the point of view of the concept proposed in the article are the results concerning the expectations of SMEs that can support the implementation of the Industry 4.0 concept (Figure 3):

- Professional IT systems (platforms) to support networking of enterprises (80% of respondents).
- Business models dedicated for network cooperation of SMEs (72% of respondents).
- Training on how to implement Industry 4.0 (67% respondents).
- European programs supporting the Industry 4.0 technologies' implementation (55% respondents).

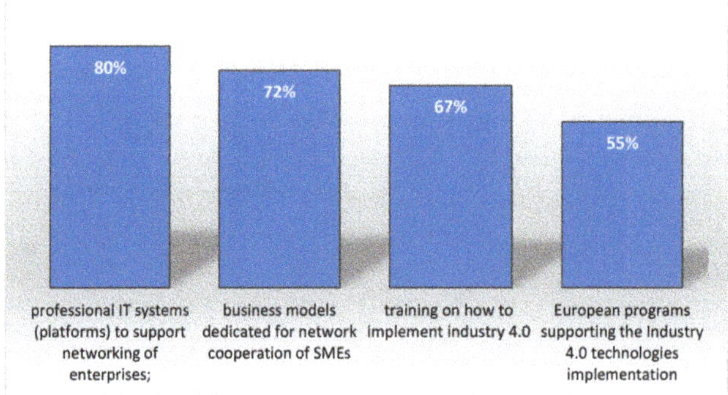

Figure 3. Expectations of SMEs that can support the implementation of the Industry 4.0 concept. Source: own elaboration.

They also indicate the main concerns related to the implementation of Industry 4.0 technology (Figure 4):

- High consulting costs in the field of new technologies (78% respondents).
- Low level of return of investment (75% respondents).
- Lack of qualified employees in the field of new technologies (67% of respondents).
- Lack of knowledge about technologies dedicated to Industry 4.0 (65% respondents).
- Problems with cooperation within networking (55% respondents).
- Low level of automation and digitization of production (45% respondents).

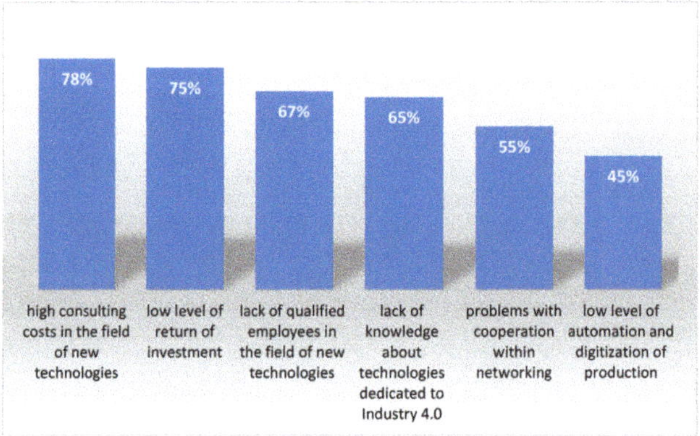

Figure 4. The main concerns of the Industry 4.0 technologies' implementation. Source: own elaboration.

Moreover, the surveyed companies, in general, are concerned about networking. The main problems with network cooperation were indicated as follows (Figure 5):

- Low level of cybersecurity for network cooperation (79% respondents).
- Lack of effective business models for the cooperation of small and medium enterprises (72% respondents).
- Logistical problems (67% of respondents).

- Disloyalty of network partners (45% of respondents).
- Difficulties in searching for partners for the networks (41% of respondents).
- Problem with co-responsibility for product quality (34% of respondents).

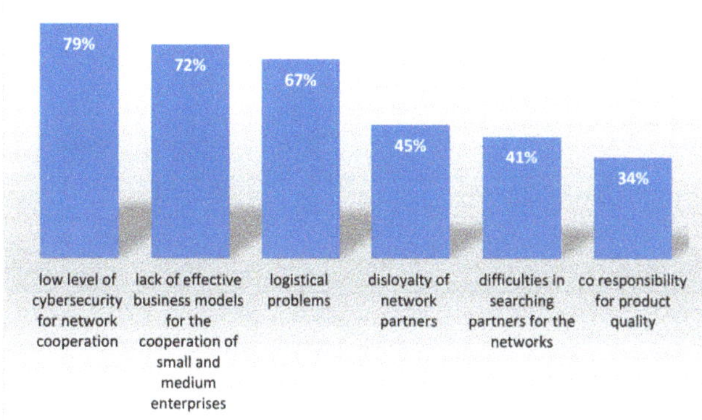

Figure 5. Main problems with network cooperation. Source: own elaboration.

5. Discussion

5.1. Business Model of the Network of Cooperating Enterprises of Industry 4.0

The ongoing development of a knowledge-based economy expressed in an intensive transfer and diffusion of innovations has a significant impact on changes in business models and business processes [49]. New forms of competitiveness and cooperation are emerging. This means that today's enterprise must abandon the patterns of the past and move from the old rules of operation envisioned for a resourceful enterprise to the formula of an intelligent, virtual and networked enterprise. The creation of network forms of cooperation is an excellent opportunity to dynamize business models within the concept of Industry 4.0 and a chance to increase competitiveness of enterprises [50]. This is confirmed by the results of bibliometric studies, which show that the technologies of Industry 4.0 enable and even facilitate and intensify the establishment of cooperation between companies within industrial networks. This is also confirmed by research among enterprises that confirms the need for cooperation throughout the implementation of the Industry 4.0 concept. Surveyed enterprises expect support in the field of European programs supporting the Industry 4.0 technologies' implementation and business models defining the terms of cooperation.

Today's economies are characterized by turbulent markets and globalization. This means for enterprises permanent changes and an intense increase in competition. There is an intensive development of technologies enabling the networking of the economy, greater integration of supply chains and cooperation of enterprises. The success of an enterprise is determined by access to real-time data and unlimited communication in a virtual environment [51]. The enterprise in the Industry 4.0 concept is seen as an intelligent module to be used in the value chain. The size of the enterprise is no longer important. In the cyber physical network, each enterprise offers various capabilities that can be used throughout the logistics chain within the Industry 4.0 concept. Modern enterprises should focus on the growth of the innovative technologies used, the level of highly qualified staff and the openness to unrestricted communication and networking. Meanwhile, the openness to unlimited communication using cloud computing, big data, and Internet of Things continues to grow rapidly. The key steps for companies to adapt to networking are shown in Figure 6.

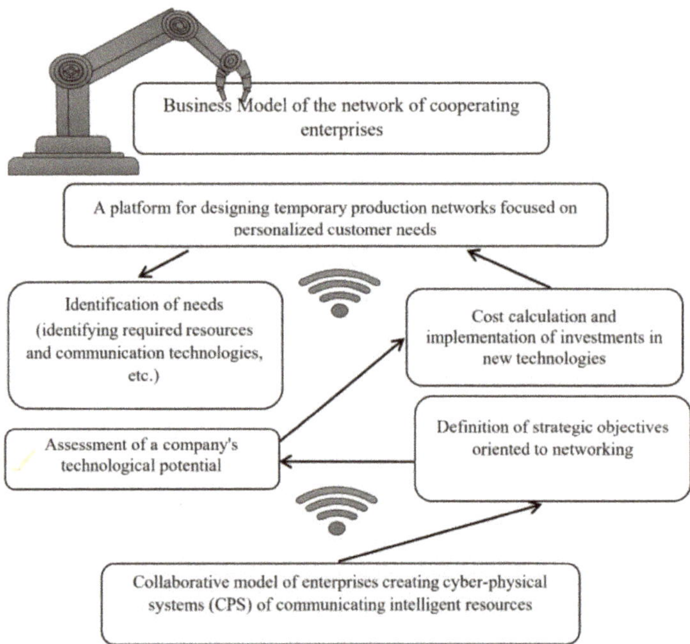

Figure 6. The key steps for companies to adapt to networking. Source: own elaboration.

The idea of a cyber-physical production network means the production order execution within shared intelligent resources of the individual network partners and communication between resources take place using real-time data and IoT [20]. An important feature of the cyber-physical network is that all network partners have access to the necessary information in real-time, regardless of the geographic location of the required resources. Thanks to the direct communication of intelligent resources, the partnership development is intensified based on combining key resources and competencies. Combining the resources of various enterprises into a network contributes to gaining a competitive advantage in the market and better orientation to the customer's needs [52].

However, in order to achieve the benefits of cooperation in the network, it is necessary to overcome the problems related to their creation, especially when the partners are enterprises from the SME sector. The process of selecting partners is complicated. Choosing a network partner requires checking, among others, partner's available production capacity, modern technology including intelligent resources, the quality level of services offered, experience, real-time communication ability, etc. [53]. Each of these areas has a major impact on the results of network creation. An equally important problem is the lack of trust, the need to invest in modern technologies and intelligent resources that are able to cooperate within the cyber-physical networks. This is especially true for the SME sector, which, due to high implementation and server costs, are only to a limited extent able to use advanced technologies. Similarly, the surveyed small and medium-sized enterprises indicate problems related to network cooperation and network forming. The most important ones include the low level of cybersecurity for network cooperation indicated by 72% of the respondents. More than half of the respondents emphasized logistical problems. Slightly less than half of the enterprises underlined difficulties in searching for partners for the networks and the problem of the disloyalty of network partners.

In the next stage, enterprises should reduce unnecessary infrastructure costs and decide on the key area of specialization in the network. This will allow for investments and, consequently, the development of narrow competencies and know-how that will be attractive to cyber-physical networks. Such an approach should guarantee a high level of use of available resources in the network and ensure the company's competitiveness in the market. The adoption of strategic goals focused on the implementation of modern technologies requires the next stage of transformation, consisting of the identification of intelligent resources and technologies required in CPS networks (German National Academy of Science and Engineering 2011). The last stage is to supplement the necessary resources and establish cooperation with the platform, the task of which is to organize temporary cyber-physical networks for the purpose of executing customer orders. Unfortunately, most of the surveyed small and medium-sized enterprises emphasized the high costs of consultancy in the field of new technologies (78% of respondents) and a low level of return on investment (75% of respondents). This can generate serious resistance to the successful implementation of network collaboration and the Industry 4.0 implementation.

The additional motivating factor to take the risk of investing in new technologies supporting the organization of cyber-physical networks is the positive experiences of enterprises, often presented in the literature. The partners cooperating in the network can offer products and services that better meet the needs of customers. The narrow specialization of enterprises, a high level of customization and the use of common resources will allow for the production of more complex and innovative products [54]. In addition, the functioning of enterprises in the network allows them to gain new experience and develop know-how. This is an advantage for both the enterprise and the network. The exchange of knowledge based on mutual relations increases the innovativeness of the offered products and services of the network [55]. Therefore, solving the problems of creating cyber-physical networks and cooperation of small and medium-sized enterprises is the basis for taking advantage of the opportunities offered by the implementation of the Industry 4.0 concept. The condition for the development of efficiently functioning networks is the development of a model of cooperation between enterprises. Created networks as CPS should ensure the collection, processing and access to data resulting from the implementation of physical material flow processes in the network. The implementation of joint production projects takes place using IoT and big data technologies used to enable unlimited communication of intelligent resources, at the same time with negligible participation of the supervising staff. The construction of the model, therefore, requires the identification of the conditions for the functioning of the enterprise in the cyber-physical production networks, the creation of temporary networks, scheduling and control of geographically dispersed intelligent resources or financial settlements of partners providing resources for production. First of all, the cyber-physical network requires an initial assessment of the technological potential, know-how, employees' competencies and resource sharing ability. This process is also focused on the production in the company [56]:

- The possibility of reducing the technological gap and ensuring technological readiness.
- Appropriate socio-technological potential.
- The ability to quickly implement innovations.

Cyber-physical production networks are characterized not only by a higher degree of functional integration but also by guaranteed easier access to the data generated by these systems (online) [57]. By networking and sharing data, enterprises are able to produce more efficiently and quickly meet customer needs. High flexibility is achieved by reducing the setup time it takes for machines to meet new requirements. Production tools can (in most cases) modify their operation on their own, adapting to new tasks—all it takes is the application of the appropriate command from the machine software [58]. This allows producers to execute small-batch production and even one or few products designed and produced as per the specification of customers, at the cost of standard mass production [56]. One of the key problems is network planning. Hence, the proposal of a concept based on building capability exchange platforms oriented towards personalized production.

5.2. The E-Business Platforms for Creating Cyber-Physical Industry Networks

One of the ways of integrating small and medium enterprises and their resources to carry out joint ventures oriented to the needs of the modern customer is building e-business platforms. The need to organize e-platforms is indicated by as many as 80% of the surveyed enterprises. In contrast, the literature review conducted shows a distinct lack of such solutions for SMEs. The professional IT systems (platforms) to support the networking of enterprises are, on the one hand, to contact the customer specifying a personalized product and, on the other hand, to integrate the resources of enterprises involved in a cyber-physical network organized for the order execution. The platform is the interface between the customer and the producer. Through the proposed offer of both products and services, the customer can specify the product and even participate in the design of a new product online. The companies around the platform have the know-how to offer design, manufacturing and transport services to temporarily established networks, to which the resources of those companies are selected that are available at the time and guarantee the timely execution of the order. Figure 7 presents the e-business platform concept.

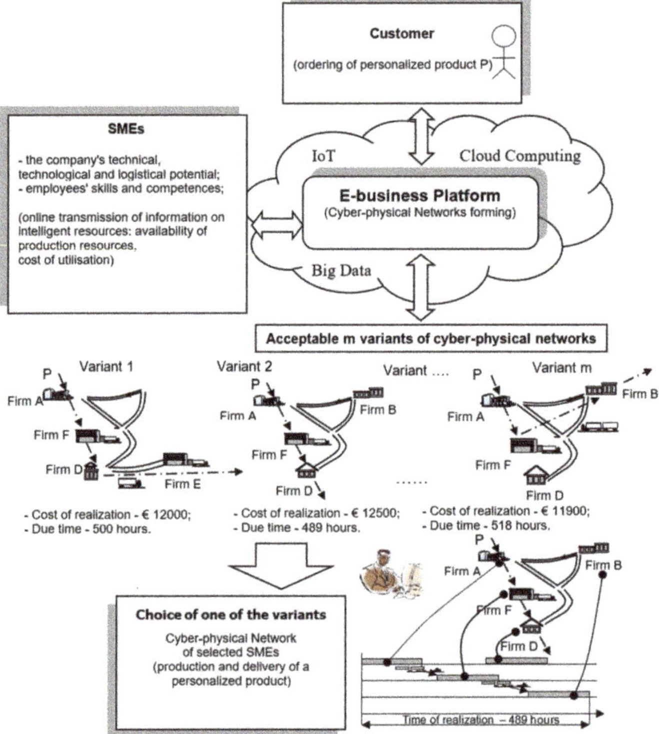

Figure 7. The e-business platform concept. Source: own elaboration.

In the Fourth Industrial Revolution era, the enterprise will be an intelligent module that is part of an integrated network of enterprises offering services to realize personalized products execution. Enterprises using the Industrial Internet of Things (IIoT) and cloud computing technologies provide real-time information about the status of intelligent resources (e.g., availability time, utilization cost, etc.) to the e-platform.

Overall, the platform is an excellent brokerage tool that, on the one hand, allows enterprises to plan tasks related to the execution of a new production order based on the specifications of the network customer, and on the other hand, allows them to collect and analyze data about the availability of enterprise resources. Based on the data provided by the resources, an appropriate algorithm based on the checking of sufficient conditions compares the generated plan for the execution of the order with the availability of resources offered by network partners. The result is a set of acceptable variants of cyber-physical networks and a detailed schedule of resource operation. From the acceptable variants, one variant is selected which has the shortest execution time or execution cost, which has a significant impact on the price. In the proposed approach, the customer decides which criterion is taken into account.

The methodology for planning a cyber-physical network is based on a sequence of checking of sufficient conditions. The proposed algorithm takes into account the constraints related to the availability of intelligent resources, the cost of their use, logistical constraints taking into account the distance between partners and the cost of transporting components in the physical material flow. In the proposed approach of cyber-physical networks' planning, three phases can be distinguished. The outline of the methodology is shown in Figure 8.

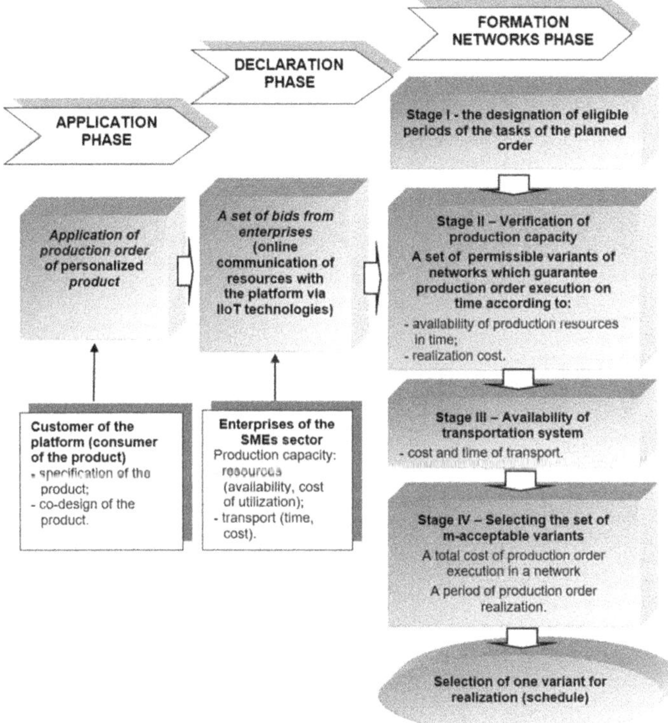

Figure 8. Outline of a platform planning methodology for cyber-physical networking. Source: own elaboration.

The first phase, called the "application phase", is the development of the design of the personalized product and the planning of the manufacturing operations that are associated with the production and delivery of the product to the customer. In this phase, the e-platform customer makes contact through an online e-commerce application and makes product specifications by selecting available product options and variants regarding shape, size, color or additional specific product features. In future, more advanced platform cases, the customer will be able to co-design the product through an intuitive computer-aided design (CAD) tool and create the product from scratch by selecting components from an available component library. After the product is accepted by the customer and sent for realization, the system will automatically create a plan of production operations selecting the appropriate manufacturing technology and production documentation.

In the next phase, the "declaration phase", the companies with adequate resources provide real-time information about the availability of production resources and information related to the cost of their use. The real-time transfer of information is possible by using the Industry 4.0 technologies (Industrial Internet of Things (IIoT), cloud computing and big data).

In the third phase, network variants are formed on the basis of the selection of resources for production operations, taking into account the succession of operations and logistical aspects related to the transportation of all materials and components during the production process for each of the acceptable network variants. At the same time, the cost of order execution is determined, which is the basis for determining the product price. As a result of this phase, a set of production networks representing a set of resources of different companies is obtained. Each variant is characterized by the time and cost of product realization. The final variant of the network is selected by the customer based on these two criteria. The price and lead time accepted by the customer means that the network is formed, and the production stages are launched in the individual enterprises of the network.

The proposed network formation approach uses a sequence of checks of sufficient conditions, the fulfilment of which guarantees the execution of the planned production order based on the analysis of data on available resources transmitted in real time from individual resources. The proposed planning approach allows for rapid prototyping of network variants. The proposed planning methodology is less labor and time consuming in contrast to the modeling and simulation methods and ad-hoc approach often used in such network planning cases. Nowadays, small and medium enterprises do not have effective methods for rapid network planning, especially in terms of partner selection and resource planning to fulfill customer-oriented orders. Hence, further development of the proposed concept seems to be a justified direction for further research in this area. The presented research results in the context of the considerations of other authors presented in the discussion confirm the correctness of the adopted hypothesis. The organization of e-business platforms of small and medium enterprises will allow the integration of enterprise resources in order to form cyber-physical networks in the conditions of the Industry 4.0 concept.

6. Conclusions

Based on the bibliometric analysis, industrial networks are often considered in scientific papers alongside Industry 4.0. This is due to the nature of the Industry 4.0 technologies' orientation towards networking the economy and the use of dispersed, intelligent resources along the entire value chain. Although the research area of SMEs is the largest, the issues of industrial networks including production networks and Industry 4.0 are rarely addressed in relation to SMEs, indicating a definite research gap.

The concept of Industry 4.0 achieves a high level of resource integration through unlimited communication of resources and integration of enterprises offering manufacturing services. Adding services to the core product offerings of enterprises to create additional value for the customer—defined as servitization—allows building better interaction between the customer and the producer. In addition, servitization makes better use of enterprise resources and provides opportunities for networking. This may particularly concern the sector of small and medium enterprises, which, in contrast to large enterprises with high development potential, see an opportunity for development in the conditions of Industry 4.0 in cooperation and narrow specialization.

Based on a conducted survey, the SME sector has indicated many expectations, e.g., need for professional IT systems (platforms) to support network cooperation, new business models and training in the area of Industry 4.0 implementation. The surveyed enterprises also indicate the key problems related to the implementation of the Industry 4.0 concept.

The main problems concern network cooperation, e.g., difficulties in searching for partners for the networks, low level of cybersecurity for network cooperation, the disloyalty of network partners, lack of business models and business platforms. The other important concerns are as follows: high consulting costs in the field of new technologies, low level of return of investment lack of qualified employees in the field of new technologies and lack of knowledge about technologies dedicated to Industry 4.0. Focusing on services for the network and development of own know-how requires solving a number of problems resulting from the lack of business models of cyber-physical networks of small and medium enterprises and methods of establishing cooperation.

The concept of an e-platform supporting cyber-physical production network prototyping proposed in the paper is a way of integrating, on the one hand, the customer, who, in a particular case, becomes a consumer of the product after its purchase, with the producer. On the other hand, the e-platform can integrate small and medium enterprises or rather resources (machines, equipment, means of transport, employees) and services within the network. The servitization and organization of e-business platforms allow increasing the degree of utilization of cooperating companies' resources, thus increasing the level of resource productivity.

By providing a platform equipped with e-commerce sales applications through a web service and creating an intuitive interface for the customer, excellent integration between the future customer and distributed enterprises is achieved. The offered ability to specify the product by selecting available options or co-designing represents a significant improvement in real-time interaction with the customer.

Also noteworthy in the proposed approach is the involvement of the customer in the final selection of the network variant, thus influencing the price and time of production order execution.

The authors' future research focuses on developing detailed algorithms for prototyping network variants based on checking sufficient conditions based on information from intelligent resources of enterprises involved in the cooperation within the networks. The next step of research is developing a prototype e-business platform for small and medium industrial enterprises.

Author Contributions: The main activities of the team of authors can be described as follows: Conceptualization, S.S., S.G.; methodology, S.S., S.G.; software, S.S., S.G.; validation, S.S., S.G.; formal analysis, S.S., S.G.; investigation, S.S., S.G.; resources, S.S., S.G.; data curation, S.S., S.G.; writing—original draft preparation, S.S., S.G.; writing—review and editing, S.S., S.G.; visualization, S.S., S.G., supervision, S.S., S.G.; funding acquisition, S.S., S.G. All authors have read and agreed to the published version of the manuscript.

Funding: This research received no external funding.

Institutional Review Board Statement: Not applicable.

Informed Consent Statement: Not applicable.

Data Availability Statement: This study analyzed publicly available datasets available in the Web of Sience and Scopus database and quantitative data from companies.

Conflicts of Interest: The authors declare no conflict of interest.

References

1. Bauernhansl, T.; Hompel, M.; Vogel-Henser, B. *Industrie 4.0 in Produkten, Automatisierung und Logistik*; Springer Fachmedien: Wiesbaden, Germany, 2014.
2. Kagermann, H.; Wahlster, W.; Helbig, J. (Eds.) *Recommendations for Implementing the Strategic Initiative Industrie 4.0: Final Report of the Industrie 4.0 Working Group. Industrie 4.0: Mit dem Internet der Dinge auf dem Weg zur 4. Industriellen Revolution*; VDINachrichten: Frankfurt, Germany, 2011.
3. Zhou, K.; Liu, T.; Zhou, L. Industry 4.0: Towards Future Industrial Opportunities and Challenges. In Proceedings of the 2015 12th International Conference on Fuzzy Systems and Knowledge Discovery (FSKD), Zhangjiajie, China, 15–17 August 2015; pp. 2147–2152.

4. Saniuk, S.; Grabowska, S.; Gajdzik, B. Personalization of Products in the Industry 4.0 Concept and Its Impact on Achieving a Higher Level of Sustainable Consumption. *Energies* **2020**, *13*, 5895. [CrossRef]

5. Holtgrewe, U. New technologies: The future and the present of work in information and communication technology. *New Technol. Work Employ* **2014**, *29*, 9–24. [CrossRef]

6. Bartosik-Purgat, M.; Ratajczak-Mrozek, M. Big Data Analysis as a Source of Companies' Competitive. Advantage: A Review. *Entrep. Bus. Econ. Rev.* **2018**, *6*, 197–215. [CrossRef]

7. Reischauer, G. Industry 4.0 as policy-driven discourse to institutionalize innovation systems in manufacturing. *Technol. Forecast. Soc. Change* **2018**, *132*, 26–33. [CrossRef]

8. Yang, Z.; Jun, M. Consumer Perception of E-Service Quality: From Internet Purchaser and Non-purchaser Perspectives. *J. Bus. Strateg.* **2008**, *25*, 59.

9. Silveira, G.; Da Borenstein, D.; Fogliatto, F.S. Mass customization: Literature review and research directions. *Int. J. Prduction Econ.* **2001**, *72*, 1–13. [CrossRef]

10. Hu, S.J. Evolving paradigms of manufacturing: From mass production to mass customization and personalization. *Procedia CIRP* **2013**, *7*, 3–8. [CrossRef]

11. Vandermerwe, S.; Rada, J. Servitization of business: Adding value by adding services. *Eur. Manag. J.* **1988**, *6*, 314–324. [CrossRef]

12. Brax, S. A manufacturer becoming service provider: Challenges and a paradox. *Manuf. Serv. Qual.* **2005**, *15*, 142–155. [CrossRef]

13. Kowalkowski, C.; Gebauer, H.; Kamp, B.; Parry, G. Servitization and deservitization: Overview, concepts, and definitions. *Ind. Mark. Manag.* **2017**, *60*, 4–10. [CrossRef]

14. Vargo, S.; Lusch, R. Service-dominant logic 2025. *Int. J. Res. Mark.* **2017**, *34*, 46–67. [CrossRef]

15. Kliment, M.; Pekarcikova, M.; Trebuna, P.; Trebuna, M. Application of TestBed 4.0 Technology within the Implementation of Industry 4.0 in Teaching Methods of Industrial Engineering as Well as Industrial Practice. *Sustainability* **2021**, *13*, 8963. [CrossRef]

16. Veza, I.; Mladineo, M.; Gjeldum, N. Managing Innovative Production Network of Smart Factories. *IFAC-PapOnLine* **2015**, *48*, 555–560. [CrossRef]

17. Camarinha-Matos, L.M.; Afsarmanesh, H. Virtual Enterprise Modeling and Support Infrastructures: Applying Multi-agent System Approaches. In Proceedings of the ECCAI Advanced Course on Artificial Intelligence; Springer: Berlin/Heidelberg, Germany, 2001; pp. 335–364.

18. Sturgeon, T.J. Modular Production Networks: A New American model of Industrial Organization. *Ind. Corp. Chang.* **2002**, *11*, 451–469. [CrossRef]

19. Villa, A. Organizing a 'network of enterprises': An object-oriented design methodology. *Comput. Integrated Manuf. Syst.* **1998**, *11*, 331–336. [CrossRef]

20. Saniuk, S.; Saniuk, A.; Cagáňová, D. Cyber Industry Networks as an environment of the Industry 4.0 implementation. *Wirel. Netw.* **2019**, *27*, 1649–1655. [CrossRef]

21. Stock, T.; Seliger, G. Opportunies of Sustainable Manufacturing in Industry 4.0. *Procedia Cirp.* **2016**, *40*, 536–541. [CrossRef]

22. Goti-Elordi, A.; De La Calle-Vicente, A.; Gil-Larrea, M.; Errasti, A.; Uradnicek, J. Application of a Business Intelligence tool within the context of Big Data in a food industry company. *Dyna* **2017**, *92*, 347–353. [CrossRef]

23. Wang, L.; Törngren, M.; Onori, M. Current status and advancement of cyber-physical systems in manufacturing. *J. Manuf. Syst.* **2015**, *37*, 517–527. [CrossRef]

24. Zoubek, M.; Poor, P.; Broum, T.; Basl, J.; Simon, M. Industry 4.0 Maturity Model Assessing Environmental Attributes of Manufacturing Company. *Appl. Sci.* **2021**, *11*, 5151. [CrossRef]

25. Vane, J.; Frantisek, K.; Basl, J. Engineering companies and their readiness for Industry 4.0. *Int. J. Product. Perfomance Manag.* **2021**, *70*, 1072–1091. [CrossRef]

26. Jazdi, N. Cyber physical systems in the context of Industry 4.0. In Proceedings of the 2014 IEEE International Conference on Automation, Quality and Testing, Robotics, Cluj-Napoca, Romania, 22–24 May 2014; pp. 1–4.

27. Grabowska, S. Smart Factories in the age of Industry 4.0. *Manag. Syst. Prod. Eng.* **2020**, *28*, 2. [CrossRef]

28. Grabowska, S.; Saniuk, S. Modern marketing for customized products under conditions of fourth industrial revolution. In *Industry 4.0. A Glocal Perspective*; Jerzy, D., Aleksandra, G., Eds.; Routledge: New York, NY, USA, 2021. [CrossRef]

29. Saniuk, S.; Grabowska, S. Challenges of business model concept of small- and medium-sized enterprise cooperation. In *Industry 4.0. A Glocal Perspective*; Jerzy, D., Aleksandra, G., Eds.; Routledge: New York, NY, USA, 2021. [CrossRef]

30. Lampel, J.; Mintzberg, H. Customizing Customization. *Sloan Manag. Rev.* **1996**, *38*, 21–30.

31. Zhou, F.; Ji, Y.; Jiao, R. Affective and cognitive design for mass personalization: Status and prospect. *J. Intell. Manuf.* **2013**, *245*, 1047–1096. [CrossRef]

32. Kumar, A. From mass customization to mass personalization: A strategic transformation. *Int. J. Flex. Manuf. Syst.* **2007**, *19*, 533–547. [CrossRef]

33. Arora, N.; Dreze, X.; Ghose, A.; Hess, J.; Iyengar, R.; Jing, B.; Joshi, Y.; Kumar, V.; Lurie, N.; Neslin, S.; et al. Putting one-to-one marketing to work: Personalization, customization, and choice. *Mark. Lett.* **2008**, *19*, 305–321. [CrossRef]

34. Młody, M. Product personalization and Industry 4.0—Assessment of the rightness of the implementation of modern technologies in the manufacturing industry from the consumers' perspective. *Ekon. Organ. Przedsiębiorstwa* **2018**, *3*, 62–67.

35. Senanayake, M.; Little, T. Mass customization: Points and extent of apparel customization. *J. Fash. Mark. Manag. Int. J.* **2010**, *14*, 282–299. [CrossRef]

36. Raddats, C.; Kowalkowski, C.; Benedettini, O.; Burton, J.; Gebauer, H. Servitization: A contemporary thematic review of four major research streams. *Ind. Mark. Manag.* **2019**, *83*, 207–223. [CrossRef]
37. Fatorachian, H.; Kazemi, H. A criticalinvestigation of Industry 4.0 in manufacturing: Theoretical operationalisation framework. *J. Prod. Plan. Control Manag. Oper.* **2018**, *29*, 633–644. [CrossRef]
38. Opresnik, D.; Taisch, M. The value of Big Data in servitization. *Int. J. Prod. Econ.* **2015**, *165*, 174–184. [CrossRef]
39. Baines, T.; Ziaee Bigdeli, A.; Bustinza, O.; Shi, V.; Baldwin, J.; Ridgway, K. Servitization: Revisiting the state-of-the-art and research priorities. *Int. J. Oper. Prod. Manag.* **2017**, *37*, 256–278. [CrossRef]
40. Baines, T.S.; Lightfoot, H.W.; Benedettini, O.; Kay, J.M. The servitization of manufacturing: A review of literature and reflection on future challenges. *J. Manuf. Technol. Manag.* **2009**, *20*, 547–567. [CrossRef]
41. Vendrell-Herrero, F.; Bustinza, O.F.; Parry, G.; Georgantzis, N. Servitization, digitization and supply chain interdependency. *Ind. Mark. Manag.* **2017**, *60*, 69–81. [CrossRef]
42. Coreynen, W.; Matthyssens, P.; Van Bockhaven, W. Boosting servitization through digitization: Pathways and dynamic resource onfigurations for manufacturers. *Ind. Mark. Manag.* **2017**, *60*, 42–53. [CrossRef]
43. Du, H.; Li, B.; Brown, M.A.; Mao, G.; Rameezdeen, R.; Chen, H. Expanding and shifting trends in carbon market research: A quantitative bibliometric study. *J. Clean Prod.* **2015**, *103*, 104–111. [CrossRef]
44. Strozzi, F.; Colicchia, C.; Creazza, A.; Noe, C. Literature review on the Smart Factory concept using bibliometric tools. *Int. J. Prod. Res.* **2017**, *55*, 6572–6591. [CrossRef]
45. Straka, M.; Malindzakova, M.; Trebuna, P.; Rosova, A.; Pekarcikova, M.; Fill, M. Application of extendsim for improvement of production logistics' efficiency. *Int. J. Simul. Model.* **2017**, *16*, 422–434. [CrossRef]
46. Trebuňa, P.; Pekarcikova, M.; Edl, M. Digital Value Stream Mapping Using the Tecnomatix Plant Simulation Software. *Int. J. Simul. Model.* **2019**, *18*, 19–32. [CrossRef]
47. Ziegler, B. *Methods for Bibliometric Analysis of Research: Renewable Energy Case Study*; Massachusetts Institute of Technology: Cambridge, MA, USA, 2009.
48. Miao, F.; Wang, G.; Jiraporn, P. Key supplier involvement in IT-enabled operations: When does it lead to improved performance? *Ind. Mark. Manag.* **2018**, *75*, 134–145. [CrossRef]
49. McAfee, A.; Brynjolfsson, E. Big data: The management revolution. *Harv. Bus. Rev.* **2012**, *90*, 60–68.
50. Birkel, H.S.; Veile, J.W.; Müller, J.M.; Hartmann, E.; Voigt, K.-I. Development of a risk framework for Industry 4.0 in the context of sustainability for established manufacturers. *Sustainability* **2019**, *11*, 384. [CrossRef]
51. Baraldi, E.; Gressetvold, E.; Harrison, D. Resource interaction in inter-organizational networks: Foundations, comparison and a research agenda. *J. Bus. Res.* **2012**, *65*, 266–276. [CrossRef]
52. Xu, L.D.; Duan, L. Big data for cyber physical systems in industry 4.0: A survey. *Enterp. Inf. Syst.* **2019**, *13*, 148–169. [CrossRef]
53. Napoleone, A.; Macchi, M.; Pozzetti, A. A review on the characteristics of cyber-physical systems for the future smart factories. *J. Manuf. Syst.* **2020**, *54*, 305–335. [CrossRef]
54. Rossit, D.A.; Tohmé, F.; Frutos, M. A data-driven scheduling approach to smart manufacturing. *J. Ind. Inf. Integr.* **2019**, *15*, 69–79. [CrossRef]
55. Schuh, G.; Potente, T.; Wesch-Potente, C.; Hauptvogel, A. Sustainable Increase of Overhead Productivity due to Cyber-Physical-Systems. *Procedia CIRP* **2013**, *19*, 51–56. [CrossRef]
56. Cheng, G.; Liu, L.; Qiang, X.; Liu, Y. Industry 4.0 Development and Application of Intelligent Manufacturing. In Proceedings of the 2016 International Conference on Information System and Artificial Intelligence (ISAI), Hong Kong, China, 24–26 June 2016; pp. 407–410. [CrossRef]
57. Lee, J.; Bagheri, B.; Kao, H. Research Letters: A Cyber-Physical Systems architecture for Industry 4.0-based manufacturing systems. *Manuf. Lett.* **2016**, *3*, 18–23. [CrossRef]
58. Oks, S.J.; Fritzsche, A.; Möslein, K.M. An Application Map for Industrial Cyber-Physical Systems. *Ind. Internet Things* **2017**, *62*, 21–45.

energies

Article

Optimization of Industry 4.0 Implementation Selection Process towards Enhancement of a Manual Assembly Line

Amanda Aljinović, Nikola Gjeldum, Boženko Bilić and Marko Mladineo *

Faculty of Electrical Engineering, Mechanical Engineering and Naval Architecture, University of Split,
R. Boskovica 32, 21 000 Split, Croatia; amaljino@fesb.hr (A.A.); ngjeldum@fesb.hr (N.G.); bbilic@fesb.hr (B.B.)
* Correspondence: mmladine@fesb.hr; Tel.: +385-21-305-939

Abstract: Last year's developments are characterized by a dramatic drop in customer demand leading to stiff competition and more challenges that each enterprise needs to cope with in a globalized market. Production in low-mix/high-volume batches is replaced with low-volume/high-variety production, which demands excessive information flow throughout production facilities. To cope with the excessive information flow, this production paradigm requires the integration of new advanced technology within production that enables the transformation of production towards smart production, i.e., towards Industry 4.0. The procedure that helps the decision-makers to select the most appropriate I4.0 technology to integrate within the current assembly line considering the expected outcomes of KPIs are not significantly been the subject of the research in the literature. Therefore, this research proposes a conceptual procedure that focus on the current state of the individual assembly line and proposes the technology to implement. The proposed solution is aligned with the expected strategic goals of the company since procedure takes into consideration value from the end-user perspective, current production plans, scheduling, throughput, and other relevant manufacturing metrics. The validation of the method was conducted on a real assembly line. The results of the validation study emphasize the importance of the individual approach for each assembly line since the preferences of the user as well as his diversified needs and possibilities affect the optimal technology selection.

Keywords: Industry 4.0; assembly line; information flow; decision making

Citation: Aljinović, A.; Gjeldum, N.;
Bilić, B.; Mladineo, M. Optimization
of Industry 4.0 Implementation
Selection Process towards
Enhancement of a Manual Assembly
Line. *Energies* **2022**, *15*, 30.
https://doi.org/10.3390/en15010030

Academic Editor: Wen-Hsien Tsai

Received: 29 November 2021
Accepted: 17 December 2021
Published: 21 December 2021

Publisher's Note: MDPI stays neutral
with regard to jurisdictional claims in
published maps and institutional affiliations.

1. Introduction

Globalization has created considerable challenges that production-oriented enterprises need to tackle: fierce competition, short windows of market opportunity, frequent launches of products and large-scale alternations in product demand [1]. The milestone for industry surely is the last year's events developments, followed by the dramatic change in customer demands. Therefore, in order to survive the turbulences on today's market, flexibility, scalability and agility have to be primary objectives for any enterprise. In addition, high degree of specialization while maintaining flexible and fast response on customer demands is the characteristic that today's enterprise needs to own. Many manufacturing enterprises have oriented their production towards more agile production approaches instead of mass production [2], in order to take opportunities enabled by market widening, where all competitors have similar opportunities, and customer demands more customized or even personalized products.

Manufacturing, as a cornerstone of the developed nations' economy, presents a strong base for any advanced country. Besides encouraging and stimulating all the other economic sectors, manufacturing provides a wide range of different jobs, thus enabling higher standards of living. Croatia's manufacturing industry has predominant lack of competitiveness due to inherent problems and obstacles. The initial plan for last three decades—to restructure the economic subjects efficiently to become competitive enough with—global markets, failed due to some transitional problems and previous economic system anomalies. Today, low industry subjects' productivity is additionally burdened by an oversized government administration, the disproportionate number of workers, obsolete technology, and lack of digitalization. Unfortunately, insufficient education of personnel in the management field, contributes to additional lagging back when compared with competitors around the globe. The lag is emphasized in the field of manufacturing personnel's lifelong learning, apathy, and idealess of management for the implementation of advanced organizational methodologies. Individual investments in cutting-edge technologies often do not result in expected outcomes, as those are implemented without proper education and management. The majority of public enterprises disappeared in the failed and corruptive privatization process, and those that managed to survive, have undergone numerous restructuring programs to avoid liquidation. In these conditions, small and medium-sized enterprises (SME) did not have sufficient support from the government administration nor by large industrial systems. Due to all these facts, economic development has been mostly turned to the service sector, especially tourism, which is turned to be crucial mistake, considering the world crisis in 2020.

Rapid response to customer demands represents a key factor of the company's survival [3]. The switch to Reconfigurable Manufacturing System (RMS) could be a solution. RMS has three major principles [1]: it ensure adaptable production resources in order to respond to unpredictable market demands and system behavior uncertainty, it is designed to produce all members of the product family, rather than single product and its system core features should be incorporated in it components (mechanical, communications and control), as well as the system as a whole. Therefore, RMS possesses flexibility of flexible manufacturing systems and productivity of dedicated manufacturing Lines [4]. Switching from a traditional manufacturing system to RMS leads toward the so-called living factory [4] or Smart Factory [5]. Such a factory can rapidly respond to the needs of customers while maintaining low manufacturing costs and high levels of quality [1]. The Smart Factory introduces Smart Product to production, that are uniquely identifiable and traced to be located any time [6]. Knowing their own history, current status and alternative routes to achieving their target state, makes the product unique (single-item) [7,8]. Smart Factories could make single item production profitable and therefore allows individual customer requirements.

To address and overcome the current challenges of shorter product lifecycles and individual customer requirements concept Industry 4.0 (I4.0) [9–11] is introduced to manufacturing systems. I4.0 [5] focuses on the establishment of products and production processes thus emphasizing the needed transformation of today's factories into Smart Factories for the production of Smart Products that are networked in an Internet of Things (IoT). A possibly worldwide network of interconnected and uniform addresses objects that communicate via a standard protocol could be considered as IoT [12]. Therefore, IoT enables the collection of production real-time data and it's exchanging among systems within factory, machine tools, workers and even customers. Information availability presents a crucial factor that enables response, almost in a real-time manner, to the changeable market demands by the rapid adjustment of an existing manufacturing system [13].

The aim of I4.0 is to achieve monitoring and synchronization of data between the physical factory floor and cyberspace. Digitalization of the manufacturing process results with some level of cognition if the information from all connected systems are collected, mutually summarized, analyzed and used for further actions [14]. That defines the manufacturing as "smart". Based on the [15], total 10 I4.0 transformation pillars can be defined: IoT, cloud computing, cyber-physical system, big data, autonomous robots, simulation, horizontal and vertical integration, additive manufacturing, augmented reality, and blockchain. Transition to a Smart Factory affects the workers in many ways, and the workers are considered as the main value of every company. In order to successfully transform to I4.0 and therefore ensure their own survival, enterprises must continuously implement newly developed methods, technologies and skills. In order to acquire practical experience and upgrade the necessary knowledge and competences about I4.0, enterprises need to continuously invest in workers' education. Using the concept of the Learning Factory (LF) presents an emerging method of how this can be achieved.

To cope with the challenges set by the I4.0 related equipment implementation and workers' education to work with, the academic institutions worldwide and some large industrial enterprises are developing a new nearly realistic factory environment for education and research called LF. After the occurrence of I4.0, diverse academic institutions have invested in their conventional LF in order to develop and build I4.0-based LF that serves as a good polygon for the transformation of the new concepts of I4.0 toward their students and industry workers. If managed properly, LFs enables students and workers introduction to the latest advancements in I4.0 and achievable results of the manufacturing and Information and Communication Technology (ICT) integration [16]. A realistic environment to test the basic engineering principles of I4.0 equipment could be achieved in LFs environment [17]. Within LFs, process improvement could be validated without cost pressure that appears in the real industrial environment. Moreover, LFs give the students and workers the ability to understand the behavior of real production systems, and the opportunity to apply different improvement scenarios in order to explore possible outcomes [18]. In order to gain workers' knowledge faster, LFs have to offer insight into innovations that can be applied in a real industry environment in order to get practical experience and skills timely [16,19]. So, they become prepared and skilled for operation in the I4.0 environment or to upgrade the conventional production systems to become smart production systems. It is expected that the number of LFs will increase in the upcoming years due to the increasing demand for better forms of learning [20].

Collaboration between academic institutions and industry is crucial and can be established by LFs. Producing knowledge through research, diffusing knowledge through education as well as using and applying knowledge through innovation is the appropriate approach, known as "the knowledge triangle". Academic institutions and industrial training facilities have to continuously adapt and enhance their education concepts and methods, in order to conform to future job profiles and related competency requirements. As innovative learning environments, LFs mostly act in an interdisciplinary manner, which has proven to be an effective concept addressing these challenges [21]. The LFs' mission is to integrate design, manufacturing and business realities into the engineering curriculum, especially by introducing hands-on experience in design, manufacturing, and product realization [22]. This is accomplished by providing the balance between engineering science and engineering practice [23,24].

Experimental research for this article was performed in the LF at the Faculty of Electrical Engineering, Mechanical Engineering and Naval Architecture, University of Split, Croatia [25]. This LF is called the Lean Learning Factory (LLF), as its initial goal, during its development, was to teach lean tools and methods by hands-on simulations. Lean management tools and methods detect any activity which does not add value to the product as waste and remove those from the production process [26]. Together with optimizing activities that are necessary, but do not add value (non-value added activities), the ultimate goal is to reduce costs, needed resources, and total production time. I4.0 enables the computerization of the so-called third industrial revolution, thus making the manufacturing process smarter, more effective and productive. In the literature, it can be found that the lean concept acts as a basis and prerequisite for I4.0 implementation [27]. In [28], another author emphasized the importance of the lean management implementation to a certain level for adopting any new methodologies, including I4.0. Therefore, upgrading LLF from pure lean method training to the I4.0 demonstration and training facility could be considered an appropriate development path. Besides research and development of some I4.0 related equipment, for laboratory usage, demonstration of I4.0 related equipment integration in the production system will remain the main focus in further LLF development. It is found the recently built smart LFs worldwide train the students and workers on how the real I4.0-based smart factory operates rather than teaching the core concepts of I4.0

related equipment and the transformation process to achieve I4.0-based smart factory in the first place [29].

Through equipment used in real industry and specially developed equipment that imitates real industry equipment, supplemented with specialized equipment for learning purposes, LF is established for research, development, demonstration and knowledge transfer to the economy [17]. The knowledge is transferred by training students in the last stages of graduate and postgraduate studies, as well as industry workers, that are trained in LLF on a different basis. Some annual conferences held at the University of Split, FESB, have workshops and activities in LLF. Individual projects with industry partners on enhancing workers' skills are mostly held in LLF to convince participants of tools and methods purposefulness. Internationally funded projects that are mainly focused on the enhancement of fundamental science achievements on scientific and academic institutions, enables the introduction of developed methods and tools to partners from industry. LLF is therefore again an appropriate facility to demonstrate project outcomes, to industry partners' management and to industry workers. The aforementioned activities in LLF require both specially developed didactic equipment and real industry equipment.

An assembly line (AL) presents a final step of the production system. It consists of consecutive workstations. On each workstation operators, i.e., workers, sequentially perform a subset of assigned assembly tasks in order to create the required subassembly or finished product within a given time range. The assembly line, called the traditional assembly line, is first introduced by Henry Ford as a tool that was supposed to fulfill the growing customer needs for the single product model while ensuring productivity growth together with total production cost reduction [30]. However, in today's industrial environment where change and unpredictability of market demand have become a constant, traditional ALs are no longer convenient. Therefore, new flexible assembly systems are requisite to deal with the required high product variety. The era of I4.0 and its related technology provides the opportunity for assembly systems to adapt to the challenges of today's global marketplace. If applied correctly, I4.0 technology can significantly contribute to higher flexibility, robustness and productivity of the AL, as well as product variety and traceability [31]. Furthermore, in order to maintain the high flexibility of the system, workforce adaptability to changes are essential. Therefore, humans still play a prime role in ALs [32], especially when it comes to assembling complex mechanical parts such as gearboxes [33]. Manual assembly is often paper-based and contains a huge amount of information about the product components of which a certain amount of information may be unnecessary and redundant [34]. Due to the variety of products and their dynamic production in today manufacturing that is characterized by low-volume/high variety of product, information that the workers must process to finished their tasks increase [35]. It means that the workers' are confronted with additional effort during their task execution that can affect their ability to comprehend complex assembly relations and consequently, increase the workers' tendency for errors. From this perspective, technologies adopted from I4.0 can provide aid and support to humans, i.e., workers, to execute the assembly task in the most effective way and consequently significantly affect the assembly system's improvement [32,36]. This is particularly true for technologies such as digital instructions, cobots, radio-frequency identification (RFID) technology and other I4.0 advanced technologies that can be deployed to improve humans' abilities.

The AL developed within the LLF seek to mimic the real industrial world as close as possible. It implies that ALs can be designed bearing in mind the utilization of equipment that is used in real industrial plants in order to produce an actual product (e.g., the gearbox), equipment such as real hand-operated tools and implements, conveyors, supermarkets, etc. Therefore, case studies applied in them can be contemplated as valuable case studies for industry [37]. Didactic games used for assembly simulation process within LLF at the University of Split entails assembly process of toy trucks and toy formulas and reworked simulation game "Lego flowcar®". In the first case where the assembly process of toys is considered, the AL consists of four workstations. Each participant has to map the assembly process according to VDI 2860 standard. When the mapping process is over, the participant is encouraged to analyze and suggest improvement of the assembly process in order to reduce the total assembly time, as well as to propose improvements in assembly line design that enables easier adjustment of assembly to automation. The second didactic game implies the usage of methods associated with warehousing and logistics systems. This simulation game includes also learning methods for workload balancing on assembly stations. Since the observed didactic games involve the assembly of toys, from the perspective of the participants, these games are not taken seriously. In order to convince the participants that the application of the methods used in didactic games executes similar results as in the assembly of real products, additional efforts were made and a gearbox assembly line was developed. Car gearboxes originate from two car models and they are produced at the factory "Zastava Automobiles". Two versions of gearbox cases, together with the high variety of different components that can be fitted, results in more than 20 diverse final goods. The product AL comprises five different workstations where the elements of ICT are installed. Therefore, the gearbox assembly line will be used as a case study for this paper.

The paper is organized in the following way: motivation and research gap are presented in Section 2; framework of proposed procedure for the evaluation of the most appropriate I4.0 technology to implement is presented in Section 3; assembly line that is used as a validation study, problem definition and implementation of the proposed decision support system is presented in Section 4; discussion and conclusion are presented in the last Section.

2. Motivation for the Proposed Procedure

I4.0 technologies are new ways that can facilitate and assist human operators during the execution of their tasks, enabling them cognitive and physical support as well as a safe environment since man is still a vital factor in manufacturing [32,38]. The introduction of new technologies within the AL must be accompanied by progress in terms of system performances, operators' well-being, and economic outcomes. Therefore, an analysis of the circumstances under which it is worthwhile to introduce new technologies is necessary [39].

To the author's knowledge, most of the reported research in the literature focused on the investigation of limitations/possibilities of the implementation of new specific I4.0 technology within manual AL, during which they focus only on one technology as a possibility. For example, one of the topics that authors often deal with is the exploration of the possibilities of applying human-robot collaboration (HRC) within the AL through the distribution of tasks between these two subjects. In these papers, the justification for the introduction of the HRC or fully automated robotic assembly in regards to traditional manual assembly is examined through the comparison of these three cases in terms of important assembly measurable performances such as total task time, batch sizes, throughput, production cost, total time etc. [37,40–43]. Uva et al. [44], Horejsi et al. [45] and Mourtzis et al. [38] investigate applicability and the effectiveness of augmented reality applications compared to traditional method (paper manual instructions). Effectiveness was investigated in terms of comparison of overall execution task time, mental demand, physical demand, errors, etc. In the paper of Wolfartsberger et al. [46], the authors point out that many technologies are not yet at a level that they can be implemented in practice, therefore they gave a review of the current technologies that support manuals assembly activities and a review on the future perspective of these technologies. Observed technologies that support assembly tasks with respect to their practical implementation in companies are HRC and instructive assistance systems (mixed and virtual reality). Hou et al. [34,47] focus on the benefits that digital instructions bring compared to paper based instruction in terms of assembly time and number of error. Yoo et al. [48] investigate the important factors for cloud computing adoption, while Marinho et al. [49] proposed decision support for I4.0 that guides decision makers in adoption of cloud enterprise resource planning. Majdzik et al. [50] proposed approach that focused on the concurrency and synchronization problem between assembly line and AGV.

Papers that are focused on the comparison of different technologies that can support assembly activities of a certain assembly line are not significantly represented in the literature. The selected technology for implementation in the assembly line should consider the economic, spatial, and cost-effective aspects and need to be aligned with the expected strategic goals of the company. A step towards that was found in the research of Peron et al. [36]. In this research, the authors investigate the possibility of applying two I4.0 technologies that assist in assembly activities. The authors proposed a conceptual decision support model for implementing cobots and digital instructions within the AL observing the cost-effectiveness of their use considering the time of task execution, throughput, cost of equipment and labor cost of workers. The framework provides guidelines that, based on the observed parameter levels, narrow the possibilities for further option configuration consideration. However, the proposed approach was developed taking into account only one product model, it means that the impact of product diversity, which is a characteristic of today's production, was neglected. Moreover, the conducted validation study showed a mismatch between the experimental solution and the solution obtained by applying this approach.

Therefore, to the best of the author's knowledge, the procedure that helps the decision-makers to select the most appropriate I4.0 technology to integrate within the current assembly line considering the expected outcomes of KPIs are not significantly been the subject in the literature. It can be noticed that the majority of the researches that deal with some kind of optimization, used only objective measurable data neglecting managers' perspectives, i.e., development strategy and uniqueness of each individual company. In accordance with that, the alignment of expected benefits with the overall strategic goal is omitted. Companies differently cope with the introduction of new technologies depending on the technological and management abilities they own. In order to achieve maximum benefits of new technologies, each company needs to better comprehend and focus on proper technologies and their applicability in their unique environment. In fact, a comprehensive and respective tool has not yet been developed that estimates the benefits of I4.0 technology and proposes the most appropriate technology, having in mind the individual needs and possibilities of the company. Therefore, the need for a decision support system that guides companies in their path of choosing the most appropriate technology, i.e., technology from which they will benefit the most, is noticed. This support system has to include both, objective measurable data and a dose of subjectivity in the decision-making process along with the criteria interdependence. The importance of inclusion of a dose of subjectivity in the decision processes during the transformation of the manufacturing towards I4.0 is also emphasized in the research by Erdogan et al. [51] where authors conclude that "leadership" is the most important criteria for finding the best strategy to transfer to I4.0. Considering the role and importance of advanced technology as enablers of I4.0 as well as the diversity of strategic aims and needs of different companies, a framework for I4.0 technology selection within the assembly line is proposed in this research. The proposed framework leads to the selection of the most appropriate I4.0 technology considering the individual criteria constraints and criteria interdependence along with the decision makers' preferences. Procedure is goal oriented, therefore it lead decision-makers to be more effective already in the first procedure step. In comparison to the approaches found in the literature, this procedure leads decision-makers to the selection of the most appropriate technology that is optimized according to the expected improvement of key performance indicators (KPI). The aim is to propose simple, efficient and easily understandable decision support system that can be readily applicable in the manufacturing context.

3. The Selection Process of the Most Appropriate I4.0 Technology

The procedure of selecting the most appropriate I4.0 technology that leads towards the enhancement of manual assembly is presented in this paper. Among the measurable criteria that were used to evaluate the selection process, the procedure will take into account the preferences of decision-makers thus enabling that subjectivism enters into the analysis. The existence of a dose of subjectivity in today's production is crucial. The usage of weights (priorities) for each observed criteria, can enable decision-makers to find the best compromise solution of the observed problem that is in accordance with the development strategy as well as with the limitations/capabilities of the company. To select the most appropriate I4.0 technology for the improvement of the AL, this study provides a general framework with three main steps, as presented in a schematic view in Figure 1. The proposed framework implies the usage of technologies that have the potency to improve the performances of the AL process by enabling constant interaction between an operator and his/her workstation.

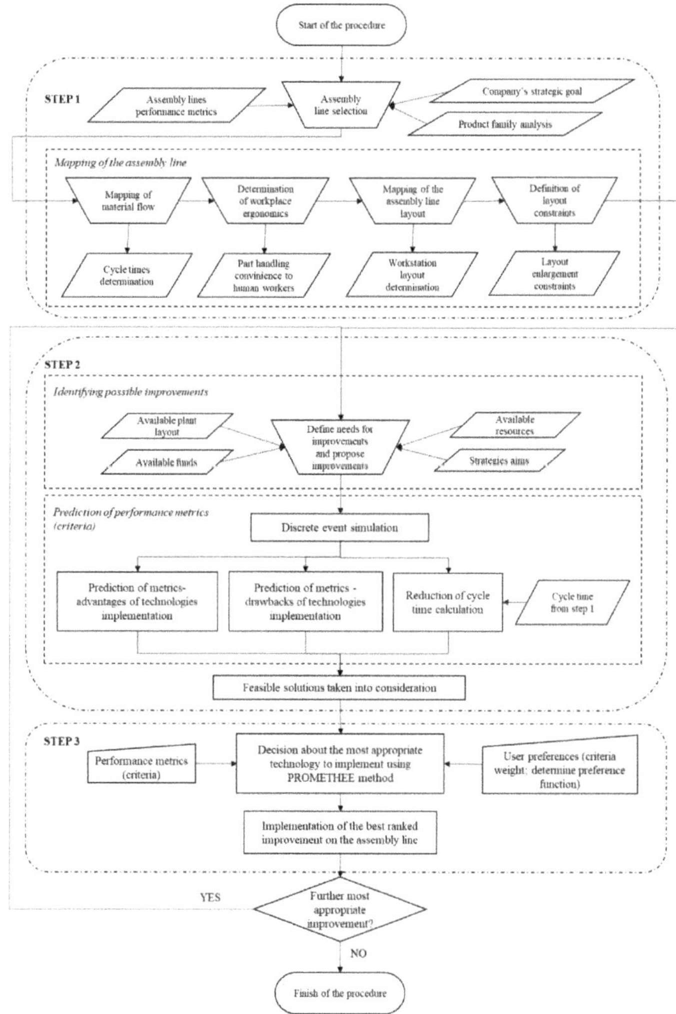

Figure 1. The proposed procedure for the evaluation of the most appropriate technology to implement.

In the first step of the procedure, according to the performance metrics and the company's strategic development aims, the AL that needs improvement is identified. AL under consideration for the improvement process by applying I4.0 technologies is confronted with problems such as unbalanced workflow, a high percentage of tasks performed by operators, unbalanced operators' utilization, high physical or mental workload of the human, high process time variations, long lead times that disrupt flexibility of AL, etc. When the line is selected, in order to evaluate the current situation and to detect issues and the possible locations for enhancements material and information flow mapping has to be done. In addition, constraints and potentiality of assembly layout should be perceived and taken into consideration, as well as other constraints that the company is confronted with. The main idea of this step is the evaluation of current performance of the AL which gives the managers, i.e., decision-makers a clear idea about performances of the line and directs them towards their strength and weaknesses in order to identify the improvement possibilities.

Prior to the I4.0 technology implementation, possible improvements of the mapped AL process condition through the commonly used methods, such as work standardization, achieving the workload balance of the assembly line, and ensuring the proper workplace ergonomics, should be considered. These previous actions may result in certain improvements in key performance indicators (KPIs) before the introduction of I4.0 technology and serve as a good foundation for further improvements.

In the second step, considering the limitations and possibilities of the AL, as well as available company's resources and strategic goals, possible improvements, in terms of different technologies that can be used, are identified. The proposed list of feasible comparable options along the whole AL serves as an input parameter for the last step of the procedure where feasible solutions are ranked according to the defined KPIs and strategic goals. Tools such as discrete event simulation can be used to predict the outcomes of the proposed technology. Outcomes refer to the KPIs that the company puts in focus as important factors through which they evaluate potential progress. To evaluate potential improvements most of the companies uses traditional KPIs such as costs, quality, line productivity, cycle time, energy consumption, throughput time, etc., but the Miqueo et al. [32] emphasize the need for new indicators that are unique for the product family, operational context and business objectives of company whose assembly system is observed. The improvement outcomes of possible technology applications should be estimated quantitatively because only measurable objective data can serve as a proper measure for optimization of the observed process.

Besides the list of feasible solutions, as input parameters in the last step of the decision support system, metrics noticed as important KPIs from prediction tools in the case of implementation of each proposed option (solution) are also used. A decision support system based on the PROMETHEE method (preference ranking organization method for enrichment evaluation) is proposed to support the multi-criteria decision-making process during the evaluation and selection of the most appropriate technology that is in line with the manufacturing and business goals and needs of the organization. PROMETHEE is an effective and significant multi-criteria decision analysis tool, it means that this method is well suited for problems where a finite set of alternatives subjected to multiple conflicting criteria has to be scalarized according to solution desirability [52,53]. The decision-makers' preferences can be set as any combination of quantitative data and the corresponding preference function of each criterion. Quantitative data are called weight (priorities) and they describe the importance of each criterion from decision makers' point of view. The sum of the weights of all the criteria is always equal to 100%. While, the preference function describes how the deviations between the assessments of two alternatives on a certain criterion should be contemplated [54]. PROMETHEE method can be briefly described through the five steps [54]. When the criteria and alternatives of a decision problem are defined, first step refers the assigning weights to each criterion and conducting the pairwise comparison in order to determine deviations between assessments of alternatives in regards to each criterion. The second step implies the selection of appropriate preference function, between six possible, for each criterion in order to model the way the decision-maker perceives the measurement scale of the criterion. The third step implies computing the global preference index:

$$\forall a, b \in A, \quad \pi(a, b) = \sum_{j=1}^{k} P_j(a, b) \times w_j, \tag{1}$$

where $\pi(a, b)$ is the global preference index of alternative a over alternative b, w_j is the weight (priority) of jth criterion, $P_j(a, b)$ is the value of preference function for jth criterion when alternative a is compared with alternative b.

The last two steps imply the computation of positive and negative preference outranking flows in order to rank all the alternatives from the best to the worst one according to the PROMETHEE I (partial ranking) and PROMETHEE II (complete ranking) ranking. Positive preference flow, $\Phi^+(a)$, measures the preference of alternative a compared to others, while negative preference flow, $\Phi^-(a)$, measures how many other alternatives are preferred to alternative a. Formula for PROMETHEE I is given in the following equations:

$$\Phi^+(a) = \frac{1}{n-1} \sum_{x \in A} \pi(a, x), \tag{2}$$

$$\Phi^-(a) = \frac{1}{n-1} \sum_{x \in A} \pi(x, a), \tag{3}$$

Formula for PROMETHEE II is given in the following formula:

$$\Phi(a) = \Phi^+(a) - \Phi^-(a), \tag{4}$$

where $\Phi(a)$ is the net outranking flow for each alternative.

The best-ranked solution, solution with the highest $\Phi(a)$ value, is the most appropriate element of I4.0 technology of which the AL benefits the most, i.e., that technology presents the best compromise solution based on the used criteria and their weights defined by the user.

The proposed procedure is implemented and validated through the study of the assembly process on the real complex product developed in the LLF environment.

4. Validation of the Developed Procedure on the Assembly Line within LLF

Assembly is the keystone manufacturing process where commodities (product parts) of all upstream manufacturing processes, from design through engineering, manufacturing and logistics are joined in order to produce and offer a functional product [55]. Assembly of complex mechanical parts, such as gearboxes, is a characteristic of the automotive industry [33]. Due to the required effort and precision of their assembly process arising from the complexity of necessary tasks, their assembly is mostly done manually. However, it can be automatized, but automation implies excessive financial investment. Some companies such as Tesla Motors have attempted excessive automation in their AL. The conclusion that emerged from their automation attempt was that humans were underestimated [56].

The validation study used in this research is the car gearbox AL situated at the LLF. Car manual gearbox is a mechanical transmission device used for torque transfer from the car engine to wheels thus enabling both, the speed change by utilizing different transmission ratios as well as reverse car drive. It consists of a huge amount of different gears, gear levers, screws, shafts, etc. The gearbox is a multi-stage product, which means that its assembly process is carried in several steps, concretely five operations steps. Each step is assigned with a certain amount of work, and each step takes place at one workstation. Each workstation contains real hand tools and supermarkets with real parts. Workstations are connected with the conveyor. In each workstation, the assembly step relates to the insertion of an individual component (part) or sets of parts that presents one gearbox subassembly element. The example of manual gearbox used in this research is given in the Figure 2.

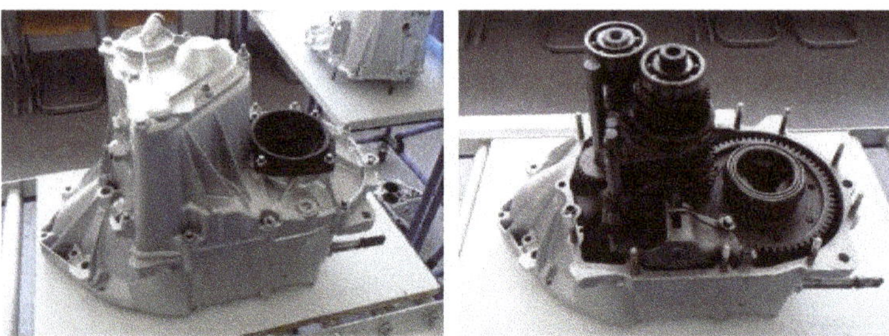

Figure 2. The assembled manual gearbox in LLF.

Gearbox AL was firstly introduced by Veza et al. [25] where the authors investigated the balancing procedure of the manual AL for only one product type using the paper instructions for workers. The further evolvement of the AL, presented by Gjeldum et al. [57], included introduction of five additional products types in order to simulate high-variety/low-volume production on the existing AL. The increase of product variety results in the inevitable increase in the number of different parts (components, subassemblies) and consequently the enormous growth of assembly information data, as well as the higher need for supermarket storage capacity. To cope with the enormous information data and to improve performances of AL, elements of I4.0 technology must be taken into account [58].

The first step towards the implementation of I4.0 technology is given by Gjeldum et al. [57] where authors presented the balancing procedure of the current state of the AL that refers to the assembly process with "the manual approach". "The manual approach" implies the usage of paper-based working instructions and manual data gathering by analysts (timing by stopwatch). Results of the balancing procedure indicate the huge discrepancy of cycle times among workstations. In order to improve work balance among workstations, further improvement of the AL is needed. The solution to this problem could be achieved with the introduction of I4.0 technology. Therefore, the presented gearbox AL represents an excellent polygon for conducting a validation study. The observed gearbox AL is shown in Figures 3 and 4.

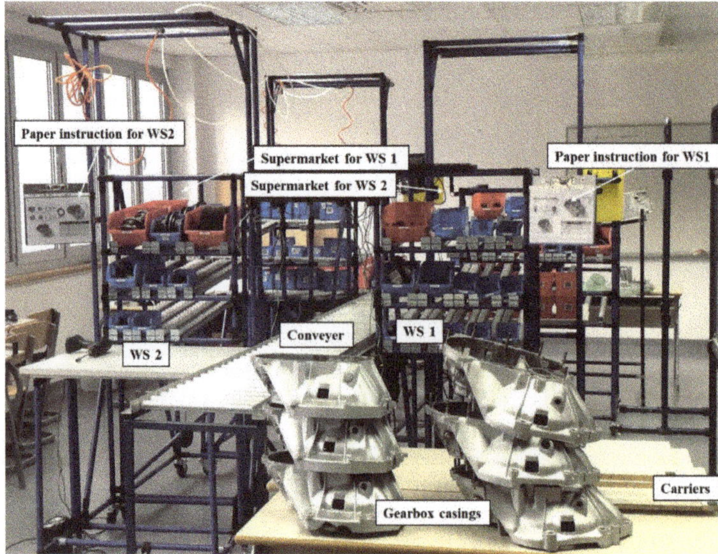

Figure 3. The gearbox assembly line in LLF.

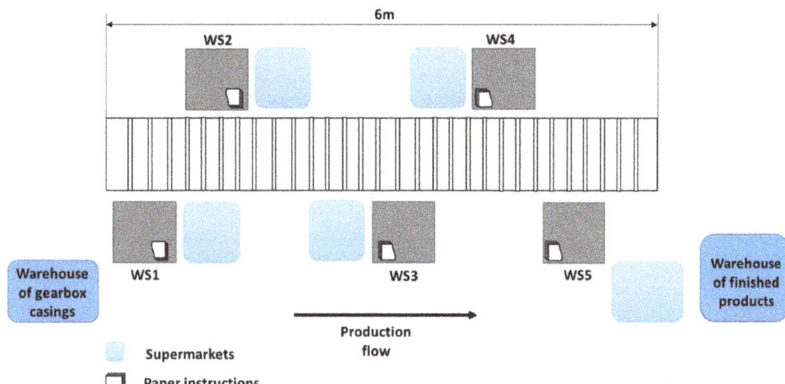

Figure 4. Top view of the gearbox assembly line.

According to the possibilities and limitations (available layout, resources and funds), as well as future strategic aims of gearbox AL development, in order to improve the yield of the assembly process, seven technologies of I4.0 are taken into consideration by presented decision support system approach. Enhancement that are taken into account are: RFID technology, Liquid Crystal Display (LCD), pick-by-light technology, augmented reality (AR), cobot, automated guided vehicle (AGV) and manipulator.

RFID technology is one of the most important technologies for automatic identification and tracking of commodities in production system. It enables precise information about the locations or states of observed goods in real-time and serves as a capstone for the establishment of the IoT within production [59]. Usage of digital instructions through LCDs instead of paper-based instructions are proved to reduce the assembly time of subassemblies/products as well as the number of errors [34,47] but only if the product is complex enough, as was stated by Syberfeldt et al. [60]. However, the meaning of the expression "complex enough" is not offered in their study. Besides the usage of digital instructions (LCD) utilization of technologies such as pick-by-light and AR also show positive impacts on assembly line performances in comparison with paper-based instructions. Pick-by-light implies the installation of a display with the light on certain shelves/boxes that contain the required parts. By light signalization, this technology guides the operators through the assembly steps, i.e., it displays screens lights up when a part has to be picked from a certain location and displays the required quantity to pick. This system is often connected with the warehouse management system. Therefore, some of the potential improvements that this technology brings are the reduction of picking time activities, as well as reduction of picking errors and operators' mental load [61,62]. When compared with paper-based instruction, AR also offers a significant improvement of the operators' performance time, error rate, cognitive (mental) load as well as minimization of the divided attention issue [38,44,63]. Noted recent research about the usage of AR technology for manufacturing purposes are papers of Zhu et al. [64], Schroeder et al. [65], Sepasgozar et al. [66] and Lalik et el. [67]. Zhu et al. [64] and Schroeder et al. [65] discussed the usage of AR technology as a tool that visualizes the Digital twin data, thus enabling the display of real-time information to the users. Sepasgozar et al. [66] presents the research about the application of the different AR technologies coupled with the Digital Twin for the virtual learning purposes while Lalik et al. [67] focus on the usage of AR technology together with the Digital Twin for the development of new system architecture for control of the industrial devices. Therefore, all of the mentioned technologies have the potential to improve the decision-making process and work procedure of the workers providing them with real-time information. Real-time information is necessary to enhance the timeliness and efficacy of the decision-making process [68].

Application of cobots within manual AL is desirable when the human is confronted with heavy loads and repetitive, tedious activities [69,70]. Human-robot collaboration (HRC) enables operators to share the same workspace with the cobots while proving the possibility to allocate tasks in a more flexible way [41]. If the tasks are assigned in an efficient way, this form of collaboration allows the system to evolve and rapidly accommodate the challenges of an increasing product variety and market volatility [71]. Tsarouchi et al. [72] and Kruger et al. [73] reported that the introduction of the cobots within the AL results in the reduction of time needed for a human operator to complete the task, higher efficiency, or the increase of human safety. On the other hand, some authors emphasize that implementation of the HRC is not always justified and may result in an increase in assembly time [69,74]. Justification and possibilities of cobot implementation within the gearbox assembly line that is observed within this research can be found in the previous paper of the authors [75]. Besides the cobot implementation, manipulator and AGV are considered in this research as technologies that coexist with the human in the workspace and reduce its physical load. AGV presents one of the most suitable and efficient technologies that can replace human work in the terms of goods' real-time supply and transportation within the factory environment. Manipulator is an electronic device developed to improve the ergonomics of the fifth assembly workstation, i.e., this device is developed to reduce operator physical effort, which occurs as a result of handling heavy components that need to be mounted (upper housing cover).

In the following subsection, the decision support system for I4.0 technology implementation is presented. The proposed approach is able to include user preferences that are associated with company constraints from an economic and practical point of view, as well as worker resource shortages.

4.1. Implementation of Decision Support System

For evaluating the use of different I4.0 technologies that can be utilized for AL development, this work proposes a decision support approach presented in Figure 1 that utilizes PROMETHEE method as a proven multi-criteria decision-making method. There are also other proven and widely used multi-criteria decision-making methods, such as: AHP, ELECTRE, and TOPSIS. However, PROMETHEE method was selected, because it was much easier to define indifference and preference thresholds of criteria as PROMETHEE method does, than to manually compare alternatives on each criterion as the AHP method does, and its approach is more convenient than the ideal and anti-ideal alternative approach of the TOPSIS method. Furthermore, PROMETHEE method is a modern version of the outranking approach that is used by ELECTRE method. By the proposed approach, seven options (alternatives) of I4.0 technologies are taken into account for the presented study, as was previously stated. These alternatives are evaluated through the four quantitative criteria: total investment cost, worker effort, workspace utilization and cycle time reduction. Each criterion is assigned a desired goal function, minimization or maximization. Two of the chosen criteria are intended to be maximized, namely "Cycle time reduction" and "Worker effort reduction". While the other two criteria are planned to be minimized. Criteria selected for this study, with their evaluation for every observed technology are shown in Figure 5.

Cycle time reduction is one of the main aims of the implementation of I4.0 technology since the total production time depends on the workstation with the highest cycle time. Information about cycle time reduction is gathered for feasible technology in relation to manual assembly. To collect this data, discrete event simulations are usually used, but in this research, the data are collected with the implementation of selected technology on a real AL that develops within LLF. The total investment cost is mainly affected by the price of the available equipment taken into consideration. Worker effort values are estimated according to the workers' experience and knowledge and they are collected through interviews with the different workers. Although, this factor can be more precisely calculated with the usage of some technologies, such as sensors or cameras, that provide information about physical worker effort according to the body gesture recognition [76], human physiology such as heart rate variability features [77], body movement and ability [78] etc. or by using approaches such as the one proposed by Blafos et al. [79]. Consequences of implementation of new technology always reflect in inevitable smaller or larger requirements of workspace layout. Since this resource is limited, it is taken into account. A presented list of criteria can be broadened and customizable to the company's strategy.

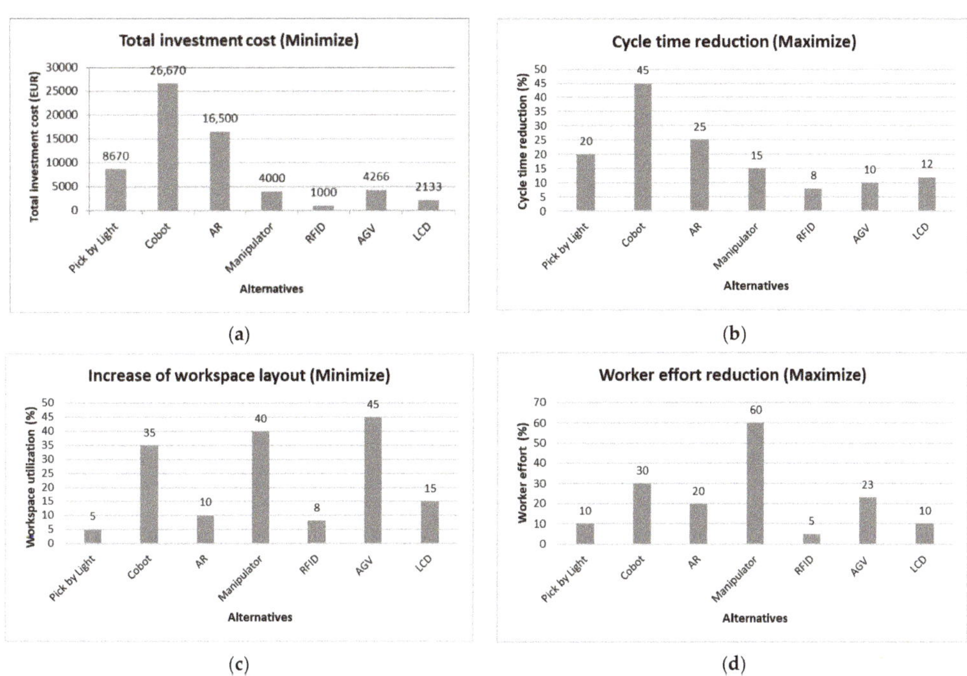

Figure 5. Criteria evaluation for each alternative: (**a**) total investment cost; (**b**) cycle time reduction; (**c**) worker effort reduction; (**d**) increase of workspace layout.

4.2. Results

The presented decision support system enables the decision-maker to express its preferences through criteria weight and preference function. For the study carried out in this research, five different criteria weights sets are observed. Each weight change refers to one scenario. The first scenario, named Scenario 0, is an alternative where each criterion has equivalent weights. The other four scenarios refer to the options in which the weight value of one criterion differs from others. In doing so, one criterion was assigned twice the weight value than the remaining criterion weights (Scenario 1–4). The weights of five observed scenarios can be seen in Figure 6.

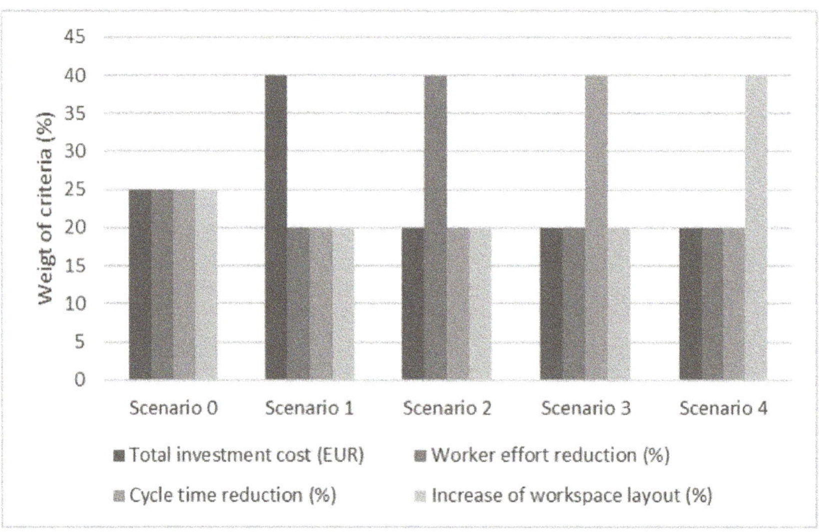

Figure 6. Different criteria weight for each scenario.

The preference function that is used for the PROMETHEE method is the linear preference function with defined indifference and preference thresholds for all criteria. Linear preference function was chosen since it is recommended for the quantitative criteria when an indifference threshold is wished to be defined, which is the situation in this study. Guidelines on how to select the appropriate preference function are given in the literature [54]. The meaning of the indifference and preference thresholds are as follows. Indifference threshold is the largest deviation that the decision-maker consider negligible, while the preference threshold is defined as the smallest deviation sufficient to generate a full preference of one alternative (option) among the other ones [54]. For example, for the presented study for the criterion "Total cost investment" the indifference threshold and preference threshold are set to 135 and 667 EUR, respectively. It means that if the difference in prices between the two observed alternatives is below EUR 135, both alternatives are equally preferred and preference of one alternative over another alternative is 0. However, if the price of alternative 1 is cheaper than alternative 2 by EUR 667 or more, alternative 1 is absolutely preferred over alternative 2. All other differences among alternatives that are between 135 and 667 EUR will result with preference between 0 and 1 of one alternative over another, according to defined linear function. Threshold values are defined as absolute number values because of simplicity and ease of understanding. The indifference and preference threshold for the remaining three criteria are given in Table 1.

Table 1. Threshold values for observed criteria.

Criterion	Indifference Threshold [1]	Preference Threshold [1]
Worker effort reduction	10	20
Cycle time reduction	5	10
Workspace layout increase	5	10

[1] Values are given in absolute amounts.

Figures 5 and 6, and Table 1 are input data for the PROMETHEE method, which is run for each scenario separately, i.e., five times. The best alternative, i.e., technologies ranking for each different scenario are compared in Figure 7 and Table 2.

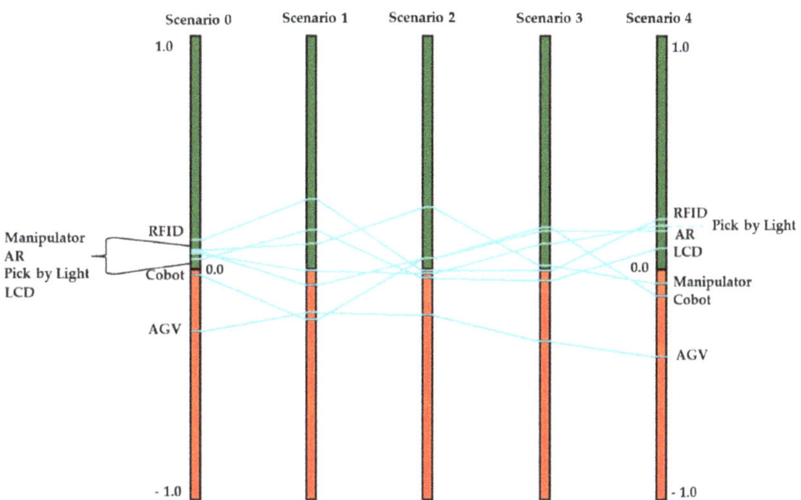

Figure 7. The rank of alternatives for each observed scenario.

Table 2. The rank of alternatives for each observed scenario.

	Scenarios (Different Weights)				
Rank of Alternatives	**Scenario 0**	**Scenario 1**	**Scenario 2**	**Scenario 3**	**Scenario 4**
1.	RFID	RFID	Manipulator	Cobot	RFID
2.	Manipulator	LCD	AR	AR	Pick by Light
3.	AR	Manipulator	Cobot	Pick by Light	AR
4.	Pick by Light	Pick by Light	RFID	Manipulator	LCD
5.	LCD	AR	Pick by Light	RFID	Manipulator
6.	Cobot	AGV	LCD	LCD	Cobot
7.	AGV	Cobot	AGV	AGV	AGV

From the alternative rank, it is clear that the RFID presents the best compromise in scenario 0 that relates to equal criterion weight because it is the cheapest option that does not increase the layout significantly. Furthermore, RFID is the best compromise solution in scenario 1 which emphasizes cost reduction. This is expected since this option is the cheapest one of all observed technology alternatives. RFID is also the best compromise solution in scenario 4 where the criterion layout increase is minimized because this technology is among the technologies that contribute the least to increasing the workspace. On the other hand, reduction of cycle time and worker effort was important in Scenario 2 and Scenario 3, respectively. Therefore, the alternative cobot resulted as the best one in Scenario 2 since this technology significantly reduces the cycle time. Because of the enormous reduction of worker effort, the manipulator option resulted as the best one for Scenario 3. It is surprising that the AGV technology in most of the scenarios come last. It can be a little bit confusing given that these types of technologies are highly used in practice due to their numerous advantages such as faster fulfillment of customer requirements and orders, reduction of production costs. One possible reason may be that these technologies could not be effectively presented through the given criteria.

The presented analysis emphasizes the importance of decision-makers preferences, i.e., user preferences, expressed through the weights (priorities) of the selected criteria for solution evaluation. Also, this analysis emphasizes the need for individual definitions of criteria depending on the needs of the enterprise. Therefore, in the process of determining the weight of the criteria, as well as during the definition and selection of important criteria that must be taken into account in the decision process, each enterprise needs to approach with great caution. It is suggested that values of weights, as well as criteria, are determined in groups when they represent the conclusion of mutual work and agreement.

5. Discussion and Conclusions

Every modern production enterprise seeks to secure its long-term survival in relentless market competition. To achieve this, keeping up with the new trends and demands set by the market, as well as keeping up with the continuous development of technology, must be the main goals of every company. Today's rapid development of technology brings the introduction of new technologies and organizational structures defined by I4.0. Therefore, constant education of workers is required in order to help enterprises quickly adapt to newly created circumstances, i.e., to help the workers to acquire knowledge and practice for further progress of their processes and organizations. The proven concept of an efficient way for worker education is the LF concept.

Since, principal features of I4.0 is the influence of technology as an accelerant that enables individualized solution, flexibility, and cost-saving in manufacturing processes [80], the proper selection of technology that brings the most benefit to the individual production system (e.g., progress within the AL of the company) is imposed for further development of enterprises. Technology should be selected having in mind the alignment of individual criteria constraints and criteria interdependence with the expected outcomes of KPIs. Therefore, in order to better adjust the selection process to individual needs of the assembly line, the proposed framework for technology selection must include an analysis of the current state of the assembly line, its limitations and possibilities as well as future strategic goals. Considering all these parameters, the selection technology procedure will guide the decision-maker to the best compromise solution. The procedure for the selection of the most appropriate technology for the development of ALs from the perspective of I4.0 is proposed in the current study. This approach takes into account the diversity of each individual enterprise production, i.e., AL, in terms of their configuration, resources, limitations, possibilities, as well as strategic aims and business policy. It means that the proposed decision support system, besides spatial and economic assets, takes into account cost-efficiency as well as alignment of expected gains to the enterprises' strategic aims. To express the real needs of the companies, the dose of subjectivity is involved in the decision-making process through the definition of criterion weight through which user preferences are expressed. This subjectivism is important, because if for example, we observe the criterion "cost" (with minimization aim) and the enterprise that has a limited budget for improvements and wants a low-cost solution, then this criterion will get a high weight value. Opposite, if the enterprise is willing to invest a lot of finance into the improvement of the production process, the criterion "cost" will have a low value. As an open approach, this procedure easily adjusts to the individual possibilities and limitations of the end-user, since it is not bound to certain methods and tools. The proposed procedure could be used to gain benefits from the I4.0 by the production managers or the CEO who is well acquainted with the assembly process.

For the purposes of this paper, a gearbox AL situated in the LF characterized by a high-variety/low-volume production type is selected to describe the proposed decision support system. Multi-criteria decision support system is based on PROMETHEE which is a proven method for ranking defined alternatives in order to select the most appropriate according to the user preferences. The highest-ranked alternative presents the best compromise from the pool of possible alternatives, considering user preferences determined *a priori*. The weights of criteria are varied in five scenarios in the gearbox AL study. For each scenario, the best option was found. Conducted validation study emphasized the impact and the importance of user preferences, as well as the need to carefully define the criteria for the evaluation of observed alternatives in the decision-making process. The selected alternative can be used immediately, but every improvement could bring certain changes. Therefore, before continuing with the following procedure iteration, assembly line balancing and a new spatial arrangement of the workstations are required to take into account. These changes could result in different values of criteria for the next iteration.

The current work contributes to the existing literature by expanding the research related to the implementation of I4.0 technology. The proposed procedure presents a good guideline for the end-user with little experience and limited resources during the technology selection process that is adjusted according to its need. It provides the rank of observed technology solutions with respect to the current state of the assembly line, its limitations, its possibilities as well as alignment of expected gains to enterprise strategic goals. The presented analysis in Section 4 emphasizes the importance of decision-makers' preferences, expressed through the criterion weights, as well as the need for criteria definition depending on the enterprise's possibilities and expected goals. Therefore, prior enterprise embarks on the implementation of this procedure, it is suggested that the values of weights, as well as the criteria definitions, are estimated in the groups. In that way, they represent the conclusion of mutual work and agreement. The proposed approach is iterative. It implies that when one selected solution is implemented on the assembly line, the whole selection procedure must be repeated in order to find the next most appropriate solution for the altered state since the values of the criteria for the next iteration could be changed. Accordingly, future research will be directed in the development of a broader decision support system. This support system will integrate the adaptive simulation models and propose the algorithm to select the roadmap of developed alternatives to avoid the iterative application of the procedure. The base of the future algorithm will be discrete event simulation coupled with the complexity I4.0 indicators that describe the changes of selected important parameters and according to them analyze the possible future alternatives. The proposed algorithm will take into account balancing of the assembly line, as well as other information related to the operational level (such as sequencing, scheduling, etc.).

Author Contributions: Conceptualization, N.G. and A.A.; methodology, B.B. and N.G.; validation, N.G., M.M. and A.A.; investigation, A.A. and N.G.; writing—original draft preparation, A.A.; writing—review and editing, N.G.; visualization, A.A.; supervision, M.M. and B.B.; funding acquisition, B.B. All authors have read and agreed to the published version of the manuscript.

Funding: Publishing of this research was funded by MZO-VIF INTEMON project.

Institutional Review Board Statement: Not applicable.

Informed Consent Statement: Not applicable.

Conflicts of Interest: The authors declare no conflict of interest.

References

1. Koren, Y. *The Global Manufacturing Revolution: Product-Process-Business Integration and Reconfigurable Systems*; John Wiley & Sons: New York, NY, USA, 2010.
2. Pasek, Z.J.; Koren, Y.; Segall, S. Manufacturing in a global context: A graduate course on agile, reconfigurable manufacturing. *Int. J. Eng. Educ.* **2004**, *20*, 742–753.
3. Hozdić, E. Smart factory for industry 4.0: A review. *Int. J. Mod. Manuf. Technol.* **2015**, *VII*, 28–35.
4. Koren, Y.; Shpitalni, M. Design of reconfigurable manufacturing systems. *J. Manuf. Syst.* **2010**, *29*, 130–141. [CrossRef]
5. Kagermann, H.; Wahlster, W.; Helbig, J. *Recommendations for Implementing the Strategic Initiative INDUSTRIE 4.0*; Heilmeyer und Sernau: Berlin, Germany, 2013.
6. Saniuk, S.; Grabowska, S. The concept of cyber-physical networks of small and medium enterprises under personalized manufacturing. *Energies* **2021**, *14*, 5273. [CrossRef]
7. Uhlemann, T.H.J.; Schock, C.; Lehmann, C.; Freiberger, S.; Steinhilper, R. The Digital Twin: Demonstrating the Potential of Real Time Data Acquisition in Production Systems. *Procedia Manuf.* **2017**, *9*, 113–120. [CrossRef]
8. Saniuk, S.; Grabowska, S.; Gajdzik, B.Z. Personalization of products in the industry 4.0 concept and its impact on achieving a higher level of sustainable consumption. *Energies* **2020**, *13*, 5895. [CrossRef]
9. Thoben, K.D.; Wiesner, S.A.; Wuest, T. "Industrie 4.0" and smart manufacturing-a review of research issues and application examples. *Int. J. Autom. Technol.* **2017**, *11*, 4–16. [CrossRef]
10. Andulkar, M.; Le, D.T.; Berger, U. A multi-case study on Industry 4.0 for SME's in Brandenburg, Germany. In Proceedings of the Annual Hawaii International Conference on System Sciences, Hilton Waikoloa Village, HI, USA, 3–6 January 2018; pp. 4544–4553.

11. Weyer, S.; Schmitt, M.; Ohmer, M.; Gorecky, D. Towards industry 4.0—Standardization as the crucial challenge for highly modular, multi-vendor production systems. In *IFAC-PapersOnLine*; Elsevier Ltd.: Amsterdam, The Netherlands, 2015; Volume 28, pp. 579–584.

12. Welbourne, E.; Battle, L.; Cole, G.; Gould, K.; Rector, K.; Raymer, S.; Balazinska, M.; Borriello, G. Building the internet of things using RFID: The RFID ecosystem experience. *IEEE Internet Comput.* **2009**, *13*, 48–55. [CrossRef]

13. Gajdzik, B.; Grabowska, S.; Saniuk, S.; Wieczorek, T. Sustainable Development and Industry 4.0: A Bibliometric Analysis Identifying Key Scientific Problems of the Sustainable Industry 4.0. *Energies* **2020**, *13*, 4254. [CrossRef]

14. Borowski, P.F. Digitization, Digital Twins, Blockchain, and Industry 4.0 as Elements of Management Process in Enterprises in the Energy Sector. *Energies* **2021**, *14*, 1885. [CrossRef]

15. Rüßmann, M.; Lorenz, M.; Gerbert, P.; Waldner, M.; Engel, P.; Harnisch, M.; Justus, J. *Future of Productivity and Growth in Manufacturing Industries*; Boston Consulting Group (BCG): Boston, MA, USA, 2015.

16. Abele, E.; Metternich, J.; Tisch, M.; Chryssolouris, G.; Sihn, W.; ElMaraghy, H.; Hummel, V.; Ranz, F. Learning factories for research, education, and training. *Procedia CIRP* **2015**, *32*, 1–6. [CrossRef]

17. Krückhans, B.; Morlock, F.; Prinz, C.; Freith, S.; Kreimeier, D.; Kuhlenkötter, B. Learning Factories qualify SMEs to operate a smart factory. In Proceedings of the COMA'16 Proceedings: International Conference on Competetive Manufacturing, Stellenbosch, South Africa, 27–29 January 2016; pp. 457–460.

18. Kreimeier, D.; Morlock, F.; Prinz, C.; Krückhans, B.; Bakir, D.C. Holistic learning factories—A concept to train lean management, resource efficiency as well as management and organization improvement skills. *Procedia CIRP* **2014**, *17*, 184–188. [CrossRef]

19. Prinz, C.; Morlock, F.; Freith, S.; Kreggenfeld, N.; Kreimeier, D.; Kuhlenkötter, B. Learning Factory Modules for Smart Factories in Industrie 4.0. *Procedia CIRP* **2016**, *54*, 113–118. [CrossRef]

20. Li, F.; Yang, J.; Wang, J.; Li, S.; Zheng, L. Integration of digitization trends in learning factories. *Procedia Manuf.* **2019**, *31*, 343–348. [CrossRef]

21. Abele, E.; Chryssolouris, G.; Sihn, W.; Metternich, J.; ElMaraghy, H.; Seliger, G.; Sivard, G.; ElMaraghy, W.; Hummel, V.; Tisch, M.; et al. Learning factories for future oriented research and education in manufacturing. *CIRP Ann.-Manuf. Technol.* **2017**, *66*, 803–826. [CrossRef]

22. Lamancusa, J.S.; Jorgensen, J.E.; Zayas-Castro, J.L. Learning Factory—A new approach to integrating design and manufacturing into the engineering curriculum. *J. Eng. Educ.* **1997**, *86*, 103–112. [CrossRef]

23. Rentzos, L.; Doukas, M.; Mavrikios, D.; Mourtzis, D.; Chryssolouris, G. Integrating manufacturing education with industrial practice using teaching factory paradigm: A construction equipment application. *Procedia CIRP* **2014**, *17*, 189–194. [CrossRef]

24. Wagner, U.; AlGeddawy, T.; ElMaraghy, H.; Müller, E. Product family design for changeable learning factories. *Procedia CIRP* **2014**, *17*, 195–200. [CrossRef]

25. Veza, I.; Gjeldum, N.; Mladineo, M. Lean learning factory at FESB—University of Split. *Procedia CIRP* **2015**, *32*, 132–137. [CrossRef]

26. Womack, J.P.; Jones, D.T.; Roos, D. *The Machine that Changed the World: The Story of Lean Production—Toyota's Secret Weapon in the Global Car Wars that is Now Revolutionizing World Industry*; Simon and Schuster: New York, NY, USA, 2007.

27. Dombrowski, U.; Richter, T.; Krenkel, P. Interdependencies of Industrie 4.0 & Lean Production Systems: A Use Cases Analysis. *Procedia Manuf.* **2017**, *11*, 1061–1068. [CrossRef]

28. Rossini, M.; Costa, F.; Tortorella, G.L.; Portioli-Staudacher, A. The interrelation between Industry 4.0 and lean production: An empirical study on European manufacturers. *Int. J. Adv. Manuf. Technol.* **2019**, *102*, 3963–3976. [CrossRef]

29. Salah, B.; Khan, S.; Ramadan, M.; Gjeldum, N. Integrating the concept of industry 4.0 by teaching methodology in industrial engineering curriculum. *Processes* **2020**, *8*, 1007. [CrossRef]

30. Wilson, J.M. Henry Ford vs. assembly line balancing. *Int. J. Prod. Res.* **2014**, *52*, 757–765. [CrossRef]

31. Bortolini, M.; Ferrari, E.; Gamberi, M.; Pilati, F.; Faccio, M. Assembly system design in the Industry 4.0 era: A general framework. In *IFAC-PapersOnLine*; Elsevier B.V.: Amsterdam, The Netherlands, 2017; Volume 50, pp. 5700–5705.

32. Miqueo, A.; Torralba, M.; Yagüe-Fabra, J.A. Lean manual assembly 4.0: A systematic review. *Appl. Sci.* **2020**, *10*, 8555. [CrossRef]

33. Gregor, M.; Medvecky, S. Digital Factory—Theory and Practice. In *Engineering the Future*; Dudas, L., Ed.; IntechOpen: London, UK, 2010; pp. 355–376.

34. Hou, L.; Wang, X.; Bernold, L.; Love, P.E.D. Using Animated Augmented Reality to Cognitively Guide Assembly. *J. Comput. Civ. Eng.* **2013**, *27*, 439–451. [CrossRef]

35. Cohen, Y.; Faccio, M.; Galizia, F.G.; Mora, C.; Pilati, F. Assembly system configuration through Industry 4.0 principles: The expected change in the actual paradigms. In *IFAC Paper Online*; Elsevier B.V.: Amsterdam, The Netherlands, 2017; Volume 50, pp. 14958–14963.

36. Peron, M.; Sgarbossa, F.; Strandhagen, J.O. Decision support model for implementing assistive technologies in assembly activities: A case study. *Int. J. Prod. Res.* **2020**, 1–27. [CrossRef]

37. Ranz, F.; Hummel, V.; Sihn, W. Capability-based Task Allocation in Human-robot Collaboration. *Procedia Manuf.* **2017**, *9*, 182–189. [CrossRef]

38. Mourtzis, D.; Zogopoulos, V.; Xanthi, F. Augmented reality application to support the assembly of highly customized products and to adapt to production re-scheduling. *Int. J. Adv. Manuf. Technol.* **2019**, *105*, 3899–3910. [CrossRef]

39. Battini, D.; Faccio, M.; Persona, A.; Sgarbossa, F. New methodological framework to improve productivity and ergonomics in assembly system design. *Int. J. Ind. Ergon.* **2011**, *41*, 30–42. [CrossRef]

40. Antonelli, D.; Astanin, S.; Bruno, G. Applicability of Human-Robot Collaboration to Small Batch Production. In Proceedings of the Collaboration in a Hyperconnected World. PRO-VE 2016, Porto, Portugal, 3–5 October 2016; Hamideh, A., Luis, C.-M., António, L.S., Eds.; Springer: Cham, Switzerland, 2016; Volume 1, pp. 24–32.
41. Faccio, M.; Bottin, M.; Rosati, G. Collaborative and traditional robotic assembly: A comparison model. *Int. J. Adv. Manuf. Technol.* **2019**, *102*, 1355–1372. [CrossRef]
42. Weckenborg, C.; Kieckhäfer, K.; Müller, C.; Grunewald, M.; Spengler, T.S. Balancing of assembly lines with collaborative robots. *Bus. Res.* **2020**, *13*, 93–132. [CrossRef]
43. Fowler, D.; Gurau, V.; Cox, D. Bridging the gap between automated manufacturing of fuel cell components and robotic assembly of fuel cell stacks. *Energies* **2019**, *12*, 3604. [CrossRef]
44. Uva, A.E.; Gattullo, M.; Manghisi, V.M.; Spagnulo, D.; Cascella, G.L.; Fiorentino, M. Evaluating the effectiveness of spatial augmented reality in smart manufacturing: A solution for manual working stations. *Int. J. Adv. Manuf. Technol.* **2018**, *94*, 509–521. [CrossRef]
45. Horejsi, P.; Novikov, K.; Simon, M. A smart factory in a smart city: Virtual and augmented reality in a smart assembly line. *IEEE Access* **2020**, *8*, 94330–94340. [CrossRef]
46. Wolfartsberger, J.; Haslwanter, J.; Lindorfer, R. Perspectives on Assistive Systems for Manual Assembly Tasks in Industry. *Technologies* **2019**, *7*, 12. [CrossRef]
47. Hou, L.; Wang, X.; Truijens, M. Using Augmented Reality to Facilitate Piping Assembly: An Experiment-Based Evaluation. *J. Comput. Civ. Eng.* **2015**, *29*, 05014007. [CrossRef]
48. Yoo, S.K.; Kim, B.Y. A decision-making model for adopting a cloud computing system. *Sustainability* **2018**, *10*, 2952. [CrossRef]
49. Marinho, M.; Prakash, V.; Garg, L.; Savaglio, C.; Bawa, S. Effective cloud resource utilisation in cloud erp decision-making process for industry 4.0 in the united states. *Electronics* **2021**, *10*, 959. [CrossRef]
50. Majdzik, P.; Witczak, M.; Lipiec, B.; Banaszak, Z. (IMS2019)Integrated fault-tolerant control of assembly and automated guided vehicle-based transportation layers. *Int. J. Comput. Integr. Manuf.* **2021**, 1–18. [CrossRef]
51. Erdogan, M.; Ozkan, B.; Karasan, A.; Kaya, I. Selecting the Best Strategy for Industry 4.0 Applications with a Case Study. In *Industrial Engineering in the Industry 4.0 Era*; Springer: Cham, Switzerland, 2018; pp. 109–119.
52. Rabbani, M.; Heidari, R.; Farrokhi-Asl, H. A bi-objective mixed-model assembly line sequencing problem considering customer satisfaction and customer buying behaviour. *Eng. Optim.* **2018**, *50*, 2123–2142. [CrossRef]
53. Abdullah, L.; Chan, W.; Afshari, A. Application of PROMETHEE method for green supplier selection: A comparative result based on preference functions. *J. Ind. Eng. Int.* **2019**, *15*, 271–285. [CrossRef]
54. Brans, J.-P.; Smet, Y. De PROMETHEE METHODS. In *Multiple Criteria Decision Analysis*; Springer: New York, NY, USA, 2016; pp. 187–219.
55. Hu, S.J.; Ko, J.; Weyand, L.; Elmaraghy, H.A.; Lien, T.K.; Koren, Y.; Bley, H.; Chryssolouris, G.; Nasr, N.; Shpitalni, M. Assembly system design and operations for product variety. *CIRP Ann.-Manuf. Technol.* **2011**, *60*, 715–733. [CrossRef]
56. Roof, K.T. Elon Musk Says 'Humans are Underrated', Calls Tesla's 'Excessive Automation' a 'Mistake'. Available online: https://techcrunch.com/2018/04/13/elon-musk-says-humans-are-underrated-calls-teslas-excessive-automation-a-mistake/ (accessed on 14 April 2021).
57. Gjeldum, N.; Salah, B.; Aljinovic, A.; Khan, S. Utilization of Industry 4.0 Related Equipment in Assembly Line Balancing Procedure. *Processes* **2020**, *8*, 864. [CrossRef]
58. Oztemel, E.; Gursev, S. Literature review of Industry 4.0 and related technologies. *J. Intell. Manuf.* **2020**, *31*, 127–182. [CrossRef]
59. Ma, H.; Wang, Y.; Wang, K. Automatic detection of false positive RFID readings using machine learning algorithms. *Expert Syst. Appl.* **2018**, *91*, 442–451. [CrossRef]
60. Syberfeldt, A.; Danielsson, O.; Holm, M.; Wang, L. Visual Assembling Guidance Using Augmented Reality. *Procedia Manuf.* **2015**, *1*, 98–109. [CrossRef]
61. Stockinger, C.; Steinebach, T.; Petrat, D.; Bruns, R.; Zöller, I. The effect of pick-by-light-systems on situation awareness in order picking activities. *Procedia Manuf.* **2020**, *45*, 96–101. [CrossRef]
62. De Vries, J.; De Koster, R.; Stam, D. Exploring the role of picker personality in predicting picking performance with pick by voice, pick to light and RF-terminal picking. *Int. J. Prod. Res.* **2016**, *54*, 2260–2274. [CrossRef]
63. Fiorentino, M.; Uva, A.E.; Gattullo, M.; Debernardis, S.; Monno, G. Augmented reality on large screen for interactive maintenance instructions. *Comput. Ind.* **2014**, *65*, 270–278. [CrossRef]
64. Zhu, Z.; Liu, C.; Xu, X. Visualisation of the digital twin data in manufacturing by using augmented reality. *Procedia CIRP* **2019**, *81*, 898–903. [CrossRef]
65. Schroeder, G.; Steinmetz, C.; Pereira, C.E.; Muller, I.; Garcia, N.; Espindola, D.; Rodrigues, R. Visualising the digital twin using web services and augmented reality. In Proceedings of the IEEE International Conference on Industrial Informatics (INDIN), Poitiers, France, 19–21 July 2016; IEEE: Piscataway, NJ, USA, 2016; pp. 522–527.
66. Sepasgozar, S.M.E. Digital twin and web-based virtual gaming technologies for online education: A case of construction management and engineering. *Appl. Sci.* **2020**, *10*, 4678. [CrossRef]
67. Lalik, K.; Flaga, S. A real-time distance measurement system for a digital twin using mixed reality goggles. *Sensors* **2021**, *21*, 7870. [CrossRef] [PubMed]

68. Gwon, S.H.; Oh, S.C.; Huang, N.; Hong, S.K. Advanced RFID application for a mixed-product assembly line. *Int. J. Adv. Manuf. Technol.* **2011**, *56*, 377–386. [CrossRef]
69. Djuric, A.M.; Rickli, J.L.; Urbanic, R.J. A Framework for Collaborative Robot (CoBot) Integration in Advanced Manufacturing Systems. *SAE Int. J. Mater. Manuf.* **2016**, *9*, 457–464. [CrossRef]
70. Villani, V.; Pini, F.; Leali, F.; Secchi, C. Survey on human–robot collaboration in industrial settings: Safety, intuitive interfaces and applications. *Mechatronics* **2018**, *55*, 248–266. [CrossRef]
71. ElMaraghy, H.; ElMaraghy, W. Smart Adaptable Assembly Systems. *Procedia CIRP* **2016**, *44*, 4–13. [CrossRef]
72. Tsarouchi, P.; Matthaiakis, A.S.; Makris, S.; Chryssolouris, G. On a human-robot collaboration in an assembly cell. *Int. J. Comput. Integr. Manuf.* **2017**, *30*, 580–589. [CrossRef]
73. Krüger, J.; Lien, T.K.; Verl, A. Cooperation of human and machines in assembly lines. *CIRP Ann.-Manuf. Technol.* **2009**, *58*, 628–646. [CrossRef]
74. Cherubini, A.; Passama, R.; Crosnier, A.; Lasnier, A.; Fraisse, P. Collaborative manufacturing with physical human-robot interaction. *Robot. Comput. Integr. Manuf.* **2016**, *40*, 1–13. [CrossRef]
75. Gjeldum, N.; Aljinovic, A.; Crnjac Zizic, M.; Mladineo, M. Collaborative robot task allocation on an assembly line using the decision support system. *Int. J. Comput. Integr. Manuf.* **2021**, 1–17. [CrossRef]
76. Wang, Z.; Lou, X.; Yu, Z.; Guo, B.; Zhou, X. Enabling non-invasive and real-time human-machine interactions based on wireless sensing and fog computing. *Pers. Ubiquitous Comput.* **2019**, *23*, 29–41. [CrossRef]
77. Murukesan, L.; Murugappan, M.; Iqbal, M.; Saravanan, K. Machine learning approach for sudden cardiac arrest prediction based on optimal heart rate variability features. *J. Med. Imaging Health Inform.* **2014**, *4*, 521–532. [CrossRef]
78. Clark, C.C.T.; Barnes, C.M.; Stratton, G.; McNarry, M.A.; Mackintosh, K.A.; Summers, H.D. A Review of Emerging Analytical Techniques for Objective Physical Activity Measurement in Humans. *Sport. Med.* **2017**, *47*, 439–447. [CrossRef]
79. Bláfoss, R.; Sundstrup, E.; Jakobsen, M.D.; Brandt, M.; Bay, H.; Andersen, L.L. Physical workload and bodily fatigue after work: Cross-sectional study among 5000 workers. *Eur. J. Public Health* **2019**, *29*, 837–842. [CrossRef] [PubMed]
80. Wang, R. *Deloitte's Study on Industry 4.0: Industry 4.0 Challenges and Solutions for the Digital Transformation and Use of Exponential Technologies*; Zurich Deloitte AG: Zurich, Switzerland, 2015.

Article

Multi-Criteria Decision Support System for Smart and Sustainable Machining Process

Luka Celent [1], Marko Mladineo [2,*], Nikola Gjeldum [2] and Marina Crnjac Zizic [2]

[1] School of Mechanical and Design Engineering, University of Portsmouth, Portsmouth PO1 3DJ, UK; luka.celent@port.ac.uk

[2] Faculty of Electrical Engineering, Mechanical Engineering and Naval Architecture, University of Split, Rudera Boskovica 32, 21000 Split, Croatia; ngjeldum@fesb.hr (N.G.); mcrnjac@fesb.hr (M.C.Z.)

*   Correspondence: mmladine@fesb.hr; Tel.: +385-21-305-939

Abstract: Sustainatableble development assumes the meeting of humanity's everyday needs and development goals while sustaining the ability of nature to provide the resources and ecosystem on which the economy and society depend. It means that an increase of economic benefit cannot be a single optimization problem anymore, instead, the multi-criteria approach is used with the accent on ecology and social welfare. However, it is not easy to harmonize these aims with machining, which is a well known industrial pollutant. On the other hand, new industrial paradigms such as Industry 4.0/5.0, are driving toward the smart concept that collects data from the manufacturing process and optimizes it in accordance with productivity and/or ecologic aims. In this research, the smart concept is used through the development of the multi-criteria decision support system for the selection of the optimal machining process in terms of sustainability. In the case of milling process selection, it has been demonstrated that green machining, without a multi-criteria approach, will always remain an interesting research option, but not a replacement for conventional machining. However, when applying realistic ecological and social criteria, green machining becomes a first choice imperative. The multi-criteria decision-making PROMETHEE method is used for the comparison and ranking, and validation of results is made through criteria weights sensitivity analysis. The contribution of this concept is that it could also be applied to other manufacturing processes.

Keywords: smart manufacturing; sustainable machining; decision support system; PROMETHEE method; Industry 4.0

Citation: Celent, L.; Mladineo, M.; Gjeldum, N.; Crnjac Zizic, M. Multi-Criteria Decision Support System for Smart and Sustainable Machining Process. *Energies* **2022**, *15*, 772. https://doi.org/10.3390/en15030772

Academic Editors: Andrew Kusiak and Surender Reddy Salkuti

Received: 22 December 2021
Accepted: 19 January 2022
Published: 21 January 2022

Publisher's Note: MDPI stays neutral with regard to jurisdictional claims in published maps and institutional affiliations.

1. Introduction

Machining is still a leading manufacturing process with a share of 60–80% in manufacturing industries of developed countries, and it has a share of 15% of total product costs [1]. For decades, optimization aims in manufacturing were: manufacture more, manufacture faster, and manufacture with fewer costs. However, the idea of sustainable development [2] does not see an economic benefit as the only optimization criterion. Instead, two more criteria are added—ecology and social welfare—making sustainability an intersection of economy, ecology, and social welfare. World leaders [3] and moral authorities [4] agreed that care for "our common home" is imperative, which requires immediate actions. According to this, the development of sustainable manufacturing processes, so called green machining, is becoming unavoidable for the manufacturing industry [5].

In the context of sustainability, machining processes are problematic. By generating different offensive pollutants and by-products, machining has a negative effect on ecology and health of workers. In the 1980 s and 1990 s, there were trends toward the clean production and green machining [6]. However, the economic side of green machining was problematic, because it represented a higher cost of manufacturing, therefore, some of the authors were not considering its sustainability, but just its preservation of the environment [6]. If looking from the environmental point of view, the usage of cutting fluids in machining is its major ecologic problem, however, it also increases productivity. The main purpose of cutting fluid application is to reduce heat generation at the cutting zone in order to prevent overheating of the cutting tool that could cause tool breakage and improve the surface roughness. Tool wear and surface roughness are always used as a quality indicator of a product. Thus, within machining processes, cutting fluids are used in order to increase productivity. The historical development of cutting fluids is brought forth by Byers [7], who concludes that it was not until the early 1960 s that scientists began to recognize and express concerns about the harmful effects of cutting fluids on the humans and environment.

At the beginning of the 21st century, on the global level, manufacturers were using close to 1.4 billion liters of cutting fluids, forming a significant demand for this non-renewable type of raw material [8]. The consumption of cutting fluids in the machining industry alone for the European Union (EU) stands at approximately 300 million liters per year [9]. Approximately 85% of cutting fluids used around the world are mineral-based cutting fluids, which cause significant environmental pollution throughout their lifecycle [10]. Having a limited lifespan, cutting fluids must be properly disposed of after use. Satisfactory disposal considers recycling or burning as a fuel. In the EU, only 32% of the total amount of annually used cutting fluids is disposed of in an environmentally friendly way [11]. Considering the losses of cutting fluids through evaporation, uncontrolled leakage, residual quantities on the workpiece, cutting tool, or swarf, it can be established that in these ways, almost 30% of the annual amount of cutting fluids used is taken out of the production system [12]. Cutting fluids are hazardous both on storage and disposal, requiring special treatment in order to remove the toxic components inside the cutting fluids before disposal. Disposal of cutting fluids raises a number of environmental issues, especially considering that it is one of the most complex types of industrial waste [13]. Cutting fluids are complex in their composition, and as such, they pose a significant health hazard throughout their life cycle. The American National Institute for Occupational Safety and Health estimates that annually, 1.2 million workers are exposed to the possible harmful effects of cutting fluids [14]. Among the most common diseases caused by exposure to cutting fluids, Ueno et al. [15] highlighted various skin diseases, while Mackerer [16] pointed out malignant and non-malignant diseases of the respiratory system. Other health hazards, including oil mist and oil vapor effects, bactericide effects, genotoxicity, the presence of cancerogenic substances and heavy metal particles [17], represent the use of cutting fluids as a burning issue with possible numerous short-term and long-term consequences for humans [18]. Regarding the economic aspects of using cutting fluids, Davim [19] wrote about its consumption in thousands of tons, allocation in billions of dollars, testifying to the scale of use and costs of cutting fluids. The costs associated with the use of cutting fluids can be up to 17% of the total production cost [20]. In the case of machining of hard-to-cut materials, that percentage can grow up to 30% of total production costs [21]. When taking into consideration the previously mentioned disposal of cutting fluids, the cost of disposal can be up to two to four times the cutting fluid purchasing costs [22].

Current usage of cutting-fluids in the machining process, which results in up to 30% of uncontrolled loss of the cutting fluids [12], makes machining a barrier in the path toward a circular economy. That fact represents an enormous issue, especially in the European Union, since the EU set a CO_2 reduction and circular economy as the main ecology aims for the next decades. The concept of Industry 5.0, presented by the European Commission [23,24] as an update of Industry 4.0 [25], also sees circular economy [26] as one of its key goals (Figure 1). The main contribution of Industry 4.0/5.0 in achieving these goals is in economic savings through better energy and resource management provided by big data analytics and optimization based on digital twins [27]. Nevertheless, the concept of Industry 4.0/5.0 is pervasive, affecting the product, process, and the whole production system [27].

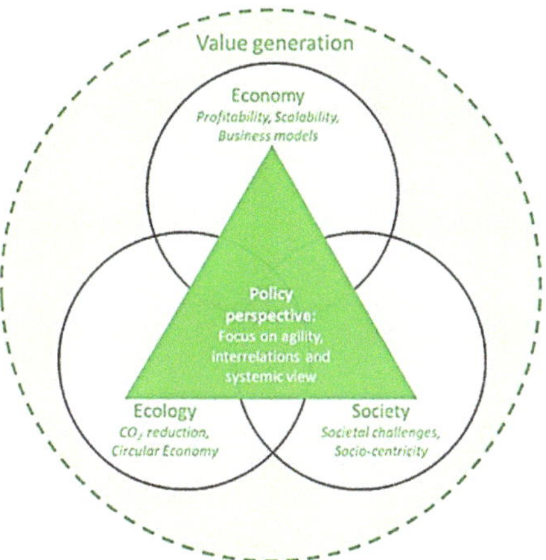

Figure 1. The main goals associated with the concept of Industry 5.0 [24].

Therefore, there is an imperative to analyze the usage of cutting-fluids in machining from the aspect of the circular economy, including social aspects. This represents a great challenge for scientists in the search for better solutions for cooling and lubrication in the machining process. These new solutions must be technically and economically viable, but also they must not represent a threat to humans or the environment. At the moment, there are several green machining techniques that add a dose of sustainability into the machining process.

1.1. Green Machining Techniques

Enactment of new legislation that will protect and enhance both human health and the environment, together with an increase in the cost of use and disposal of the cutting fluids, have led to extensive scientific research towards so-called environmentally friendly and potentially sustainable green machining [28]. In this light, the machining industry is focused on trying to achieve dry machining, as well as new alternative methods and systems such as minimum quantity lubrication (MQL) systems, compressed cold air cooling (CCA) systems, and cryogenic machining systems.

The advantages of a complete switch from the conventional use of cutting fluids to dry machining are multiple [29,30]: no negative impact on humans and the environment, reduction of variable machining costs, easier swarf recycling, no need for degreasing of the workpiece after the machining process, and in some cases longer tool life when high-speed machining some specific materials. In addition to cost-effectiveness and other mentioned practicalities of dry machining, Dinnie [31] also noted the increase in the company's positive image, which is one of the main ways to gain a lasting competitive advantage in the global market of today.

The intermediate solution towards reaching dry machining is the MQL method. The MQL method is the most used alternative to the conventional use of cutting fluids, and it can be classified as a semi-dry machining method. Reduction in power consumption and cutting fluid amounts can be made using this technique [32]. In the case of MQL, oil provides lubrication, while the effect of cooling and blowing off the swarf from the cutting zone is obtained by the constant presence of an airflow [33].

Both machining with the conventional use of cutting fluid and machining under MQL pose the problem of disposal of cutting fluids. A by-product of MQL is mist, which in the industrial environment can cause serious respiratory effects on the workers that are exposed to such substances [34]. CCA cooling and cryogenic machining systems have no such effect on workers and, as such, represent a strong alternative to any liquid coolant in machining. Initial studies used room-temperature compressed air, which proved ineffective compared to liquid coolant used during the machining process. New studies that were conducted in the conditions of the compressed cold air cooling yield the results comparable and, in some cases, even better than the use of cutting fluids. One way to perform CCA cooling is the use of specially designed systems for the preparation and distribution of compressed cold air, which requires an additional source of power [35,36]. Another way of performing CCA cooling is the use of a vortex tube which requires only the supply of a certain amount of air under pressure, enabling cold air-cooling conditions without the need for an additional source of power [37,38]. The phenomenon of the creation of cold air inside the vortex tube has not been fully explained yet, and different interpretations of this phenomenon can be found [39]. There are few studies focused on the effect of CCA cooling while machining, especially for milling operations. However, existing studies indicate great potential in the end milling of ASSAB 718HH mold steel with a hardness of 35 HRc [40], hard milling of AISI D2 cold work tool steel with a hardness of 62 HRc [31], and milling of AISI 1050 steel with a hardness of 10 HRc [41].

Cryogenic machining involves cooling with cryogenic fluids—most often liquid nitrogen—at temperatures down to $-196\ °C$ [42]. Liquid helium and carbon dioxide are also used as cryogenic fluids. Similar to CCA, cryogenic fluid is applied directly to the cutting zone in order to cool down both cutting tools and workpiece. In the case of liquid nitrogen, which accounts for 78% of the earth's atmosphere and is inert, lighter than air gas, evaporation does not pose any threat in terms of environmental pollution or any danger to workers' health [43]; the same cannot be said for carbon dioxide which pollutes the atmosphere. Some researchers are proposing a system that will use already existing carbon dioxide exhaust gases within the production plant as cryogenic fluids [44]. In that case, cryogenic machining would not contribute to additional air pollution. Furthermore, the high consumption of cryogenic fluids, as well as the cost of investment in the previously mentioned systems for supplying and applying cryogenic fluids while machining, as well as the cost of cryogenic fluids itself, ultimately increases the cost of the machining process and raises the question of the economic viability of the whole cryogenic process. It can be concluded that cryogenic processes are profitable only in specific cutting conditions using high values for cutting parameters such as cutting speed and feed [45].

In the end, it is important to mention the application of cutting fluids with the usage of biodegradable cutting fluids [32,46]. It represents an interesting and important research topic. However, it will not be considered in this research because a cutting fluid can be biodegradable, but after machining, it still needs to be processed before the process of disposal. Namely, the cutting fluid is mixed with metal chips and particles during the machining process, so it is not biodegradable in that form, but it needs to be processed lately.

In this research, three different sets of milling experiments have been used [47]: conventional milling with the application of cutting fluids (CF), dry milling (DM), and milling with the application of compressed cold air cooling (CCA). These experiments are used as input data for the developed multi-criteria decision support system, i.e., these experiments are used for the proof of concept of the developed system.

1.2. Decision Support System for Smart Manufacturing

The multi-criteria perspective of sustainable development represents a significant paradigm change not just from the industrial perspective, but from the scientific perspective as well. This means a move from single-objective optimization toward multi-objective optimization and multi-criteria decision-making. Since multi-objective optimization is used for quantitative objectives, multi-criteria decision-making, which also supports qualitative criteria, can find a wider application [48,49]. Therefore, it was suitable to apply some multi-criteria decision-making methods in this research.

The concept of a smart system implies a function of real-time sensing of the process or environment and the possibility to actuate, i.e., to control the process. However, a smart system is not necessarily autonomous—rather, it is controlled by humans most of the time, which means that the process is subordinated to human decision-making. The same principle applies to the smart manufacturing concept: sensors that are part of the industrial Internet-of-Things provide "live" data and the ability to create a digital twin of the manufacturing system. The digital twin is used for optimization and simulation of the manufacturing system in order to apply changes that will optimize the production process. However, the smart manufacturing system is not a cognitive system, therefore, decisions are made by human decision-makers.

For more than 40 years, there has been an idea of enhancing human decision-making abilities by providing them artificial decision support in the form of a computer system: the decision support system (DSS). One of the DSS pioneers—Ralph H. Sprague Jr.—developed the conceptual foundation for decision support systems in 1979. He defined the main elements of the DSS [50]: data, model, and user interface (dialogue) for interaction with humans (Figure 2).

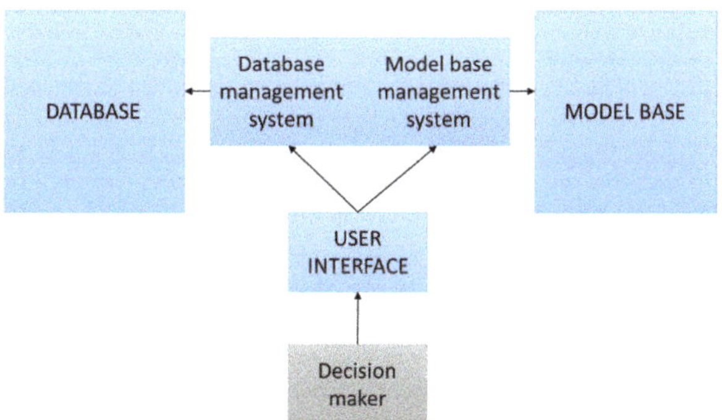

Figure 2. The main elements of the decision support system: data, model, and user interface (adapted from [50]).

In this research, the same concept of DSS (Figure 2) was used to create a multi-criteria-based decision support system for sustainable decisions in machining. The developed DSS is based on a multi-criteria decision-making PROMETHEE (Preference Ranking Organization METHod for Enrichment of Evaluations) method. As mentioned, three different sets of milling experiments—conventional milling with the application of cutting fluids (CF), dry milling (DM), and milling with the application of compressed cold air cooling (CCA) —were used to create three case studies for presentation and validation of the concept. The main research objective was to design a DSS that will ensure a proper comparison of green and conventional machining techniques, which proves that green techniques are optimal in terms of sustainability. It is a known fact that conventional machining techniques are more productive and economical than green techniques, and perhaps it will always remain that way. But, if profit is not the only optimization criterion, then green machining techniques become important. Highlighting this fact was also one of the research objectives.

Regarding existing DSS concepts for machining, most of the time, they are used to support the selection of the machining process or machine tool, but there are some authors who have focused on the selection of parameters of the machining process. Temuçin et al. [51] are using fuzzy based DSS to select machining processes aimed at non-traditional technologies. Taha and Rostam [52] are using a combination of AHP and PROMETHEE methods for the selection of machine tools, and Alberti et al. [53] are using the same problem, but they are more focused on machine characteristics and performance tests. Balazinski et al. [54] demonstrated a DSS for the selection of cutting parameters back in 1994, and it was one of the first applications of DSS in machining. Niamat et al. [55], similar to Ming et al. [56], are using multi-objective optimization to optimize electro-discharge machining process parameters. Wittbrodt and Paszek [57] developed a DSS for monitoring and forecasting tool wear based on fuzzy logic. Vidal et al. [58] are using DSS to plan milling operations by optimizing the selection of parameters.

However, there are some research regarding DSS in machining that is closer to research presented in this paper. Plaza et al. [59] aimed to stabilize cutting forces in order to achieve higher quality, but this led to a reduction of machining efficiency. Similarly, but with differences, a kind of balanced approach is presented in this paper: balancing economic benefits and machining efficiency on one side, and preserving the natural environment and human health on the other side. Shin et al. [60] and Khan et al. [5,61] have, similarly, taken sustainability criteria into account in their research. In order to do so, they are examining and optimizing the usage of energy resources during the machining process.

To conclude, the research presented in this paper contributes to the area of application of DSS in machining with the novel approach of implementing sustainability criteria into the milling process selection, which results in sustainable and green machining. It demonstrates, on real milling experiments, how to properly balance between machining efficiency and preservation of the planet for future generations. Practical contribution of the research is that developed concept could be applied to other manufacturing processes and to other human economic activities, as well.

The rest of the paper is organized as follows: in the section "Methods", the description of proposed DSS is given together with an explanation of the PROMETHEE method. The experimental data sets for three different milling techniques that will be used as a case study are also presented. The section "Results" is divided into four subsections. The first is Case 1, in which the comparison of milling techniques is based on four quantitative criteria for economic and productivity. The next subsection is Case 2, in which the comparison of milling techniques has been extended with three additional qualitative criteria for productivity. The subsection Case 3 describes the further extension of comparison of milling techniques with 2 additional qualitative criteria for ecology and society, thus rounding up the sustainability criteria. The final subsection tests the validity of the proposed approach through criteria weights stability analysis. Finally, the conclusions and suggestions for further research are given in the "Conclusions" section.

2. Methods

The starting point of this research is the definition of the machining process by defining the most important inputs and outputs. A common model for the machining process is presented in Figure 3. The inputs are the type of machining (milling, turning, etc.) and its parameters (cutting speed, feed, etc.), with the definition of the machining technique (application of cutting fluids, dry machining, etc.). The outputs can be divided into value added or productivity (surface roughness, material removal rate, etc.) and non-value added or cost (tool wear, machining cost, etc.). The optimization aim of this common machining process would be to increase productivity and decrease costs.

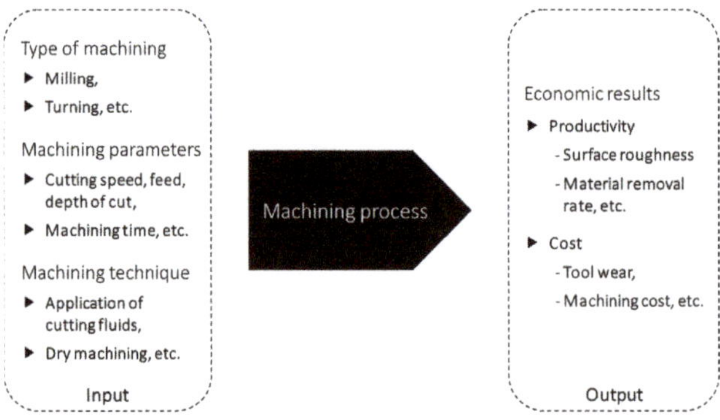

Figure 3. A common model of machining process with inputs and outputs.

It is the common approach that, however, has a limited view on negative outputs of machining, since it is only focused on economic cost. But, what about ecological costs through the negative impact on environment and social cost through negative impact on worker's health?

In this research, a model of the machining process was extended from direct outputs of machining—technical productivity and economic cost—to indirect outputs of machining—ecologic impact and threat to human health (Figure 4). Technical productivity is actually a productivity but without an economic aspect (tool wear, machining cost, etc.). Ecological impact assumes a negative impact on ecology (environment).

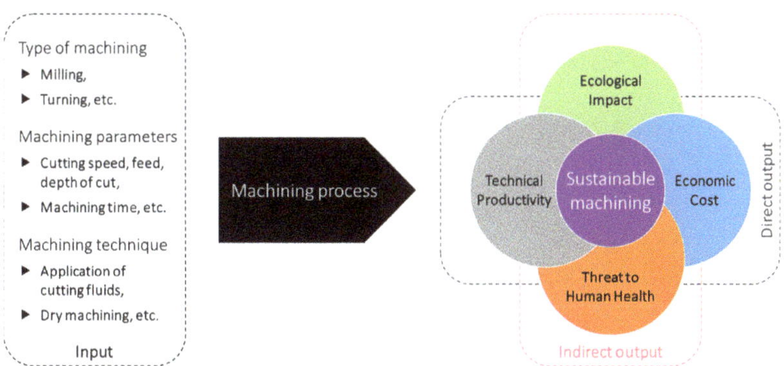

Figure 4. Proposed model of machining process with extended outputs definition.

When taking into account all proposed outputs (criteria), a machining process can be optimized in order to achieve sustainable machining. The aim of this research was to develop a decision support system for sustainable machining. The classic DSS concept was used consisting of the data, model, and user interface. An optimization model is based on multi-criteria decision-making method and uses criteria (outputs) defined in Figure 4. The data were collected through experiments and research, and the data collection process could be automated by using smart technology in the future. The user interface was a software, in this case, Visual PROMETHEE software (Ver. 1.4, Mareschal, Bruxelles, Belgium), since it has been decided to use PROMETHEE as a multi-criteria decision-making method. The described concept of the DSS is presented in Figure 5.

Today, there are many multi-criteria decision-making methods in everyday engineering practice [62]. However, some of them are better accepted and spread because of their capability to adjust to a variety of problems: AHP [63,64], ELECTRE [64], TOPSIS [64,65], and PROMETHEE [66]. A crucial issue is to decide which method is the most adequate for a particular problem, but most of the time, the outranking method, like PROMETHEE, is the most suitable choice [67]. In this research, the PROMETHEE method was selected because, for this kind of criteria, it is much easier to define indifference and preference thresholds than to manually compare alternatives on each criterion as the AHP method does. Furthermore, the PROMETHEE method better deals with the combination of qualitative and quantitative criteria than the ELECTRE method; and criteria evaluations, in this case, do not have enough large set of data to properly construct ideal and anti-ideal alternative to the TOPSIS method.

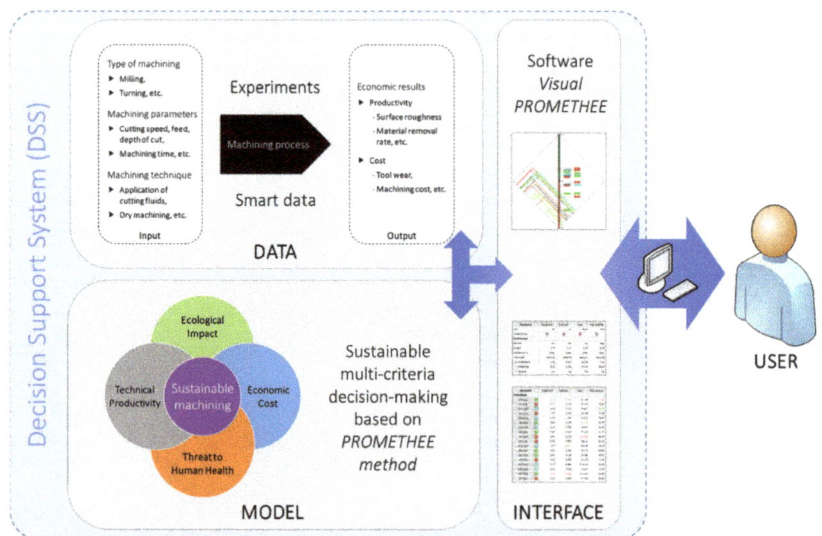

Figure 5. Proposed decision support system for smart and sustainable machining.

The PROMETHEE (Preference Ranking Organization METHod for Enrichment of Evaluations) method was developed by J.P. Brans and B. Mareschal in 1983 [68]. Input for the PROMETHEE method is a matrix consisting of a set of potential alternatives (actions) *A*, where each *a* element of *A* has its *f(a)*, which represents the evaluation of one criterion.

Method PROMETHEE I ranks actions by a partial ranking, with the following dominance flows, for the positive outranking flow [68]:

$$\Phi^+(a) = \frac{1}{n-1} \sum_{b \in A} \Pi(a, b) \tag{1}$$

and for the negative outranking flow [68]:

$$\Phi^-(a) = \frac{1}{n-1} \sum_{b \in A} \Pi(b, a) \tag{2}$$

where *a* and *b* represent the actions from a set of action *A* (during the pairwise comparison of action *a* with all other $n-1$ actions), *n* is the number of actions and Π is the aggregated preference index defined for each couple of actions.

The PROMETHEE I method gives the partial relation, and then a net outranking flow is obtained from the PROMETHEE II method, which ranks the actions by complete ranking calculating net flow [68]:

$$\Phi(a) = \Phi^+(a) - \Phi^-(a) \tag{3}$$

In the sense of priority assessment, net outranking flow represents the synthetic parameter based on defined criteria and priorities among criteria. Usually, criteria are weighted using criteria weights w_j and the usual pondering technique [68]:

$$\Pi(a, b) = \frac{\sum_{j=1}^{n} w_j P_j(a, b)}{\sum_{j=1}^{n} w_j} \tag{4}$$

where $P_j(a,b)$ represents preference of *a* over *b* for a given preference function of criterion *j*. There are six different types of preference functions, but in this research, only the linear preference function with indifference and preference thresholds were used.

As mentioned, the PROMETHEE method requires a decision matrix as an input that consists of alternatives and their criteria evaluations. In this case, alternatives are different machining techniques with different machining parameters (cutting speed, feed per tooth, depth of cut, machining time). So, five variable factors and three levels of factors were used to define the experiment set (Table 1). The experimental design proposed by Taguchi method uses the orthogonal arrays to organize the factors affecting the process, meaning the design is balanced, so that factor levels are weighted equally. Following this, every factor appears on the same number of levels, and every factor on any level will be in all combinations with other factors. Such a method allows the determination of factors that most affect the process with a minimum amount of experiments, thus saving time and resources. The levels of factors, adopted values of machining parameters correspond to the operational limits recommended by the cutting tool manufacturer together with the machine tool capabilities.

Table 1. Factors and levels used for Taguchi design of experiments [47].

Levels of Factor	Cutting Speed v_c [m/min]	Feed Per Tooth f_z [mm/zub]	Depth of Cut a_e [mm]	Machining Time t [min]	Machining Technique U_o
	A	B	C	D	E
1	100	0.05	1	10	Dry milling (DM)
2	125	0.08	1.5	16	Conventional milling with the application of cutting fluids (CF)
3	150	0.11	2	22	Milling with the application of compressed cold air cooling (CCA)

Three levels were used, since three different machining techniques—dry milling, milling with the application of cutting fluids, milling with the application of compressed cold air cooling—were used in experiments, so three levels of parameters were defined for other factors as well.

According to Taguchi's orthogonal array, a set of 18 experiments is a minimal set for this design of experiments (Table 2). These experiments were made and four outputs (results) were measured: surface roughness, tool wear, cost, and material removal.

The profilometer Mitutoyo Surftest 301 (Mitutoyo Corporation, Kawasaki, Japan) was used for surface roughness measuring. Every measurement was repeated five times and the average value was considered. Tool wear measurements were performed in accordance with the International Standard ISO 8688-1 and periodically according to the experimental plan. Tool wear of all the cutting inserts was measured and the average value was used. This was done by using the toolmaker's microscope with 100 times magnification. Material removal was calculated by using a specific equation that considers all the important cutting and experiment parameters such as axial depth of cut, radial depth of cut, feed rate, feed per tooth, mill diameter, spindle speed and number of inserts. The costs associated with each experiment included direct labor costs (cutting tool costs, cost of cutting fluid, machining time) and electricity. The cost of electricity varied between different machining techniques.

Additionally, a label was created for each experiment based on the combination of factors (Table 2). These data were used as the input matrix for the PROMETHEE method.

Table 2. Experiments' factors (input) and experiments' results for 18 experiments defined by Taguchi orthogonal array [47].

No.	Factors					Label (E ABCD)	Output (Result)			
	A	B	C	D	E		Tool Wear VB [µm]	Cost C_{so} [€]	Surface Roughnes Ra [µm]	Material Removal V_{uk} [cm3]
1.	1	1	1	1	1 (DM)	DM 1111	0.040	7.06	0.18	11.90
2.	1	2	2	2	2 (CF)	CF 1222	0.070	7.33	0.31	45.80
3.	1	3	3	3	3 (CCA)	CCA 1333	0.071	8.81	0.62	116.00
4.	2	1	1	2	2 (CF)	CF 2112	0.070	6.95	0.37	23.90
5.	2	2	2	3	3 (CCA)	CCA 2223	0.105	9.23	0.44	78.80
6.	2	3	3	1	1 (DM)	DM 2331	0.070	3.25	0.64	65.70
7.	3	1	2	1	3 (CCA)	CCA 3121	0.050	4.43	0.19	26.90
8.	3	2	3	2	1 (DM)	DM 3232	0.110	5.44	0.36	91.70
9.	3	3	1	3	2 (CF)	CF 3313	0.123	13.89	0.62	86.70
10.	1	1	3	3	2 (CF)	CF 1133	0.093	8.13	0.45	52.50
11.	1	2	1	1	3 (CCA)	CCA 1211	0.040	6.30	0.27	19.10
12.	1	3	2	2	1 (DM)	DM 1322	0.120	5.78	0.93	63.00
13.	2	1	2	3	1 (DM)	DM 2123	0.120	7.86	0.60	49.20
14.	2	2	3	1	2 (CF)	CF 2231	0.050	4.57	0.33	47.70
15.	2	3	1	2	3 (CCA)	CCA 2312	0.060	10.93	0.47	52.50
16.	3	1	3	2	3 (CCA)	CCA 3132	0.092	6.07	0.21	57.30
17.	3	2	1	3	1 (DM)	DM 3213	0.160	11.13	1.10	63.00
18.	3	3	2	1	2 (CF)	CF 3321	0.060	5.92	0.32	59.10

3. Results

It is now time to apply the PROMETHEE method on the problem of comparison of milling techniques with different machining parameters in order to select the optimal technique and parameter set. For a start, the problem consists of 18 alternatives and 4 quantitative criteria (Figure 6). Later, the problem was extended with more criteria.

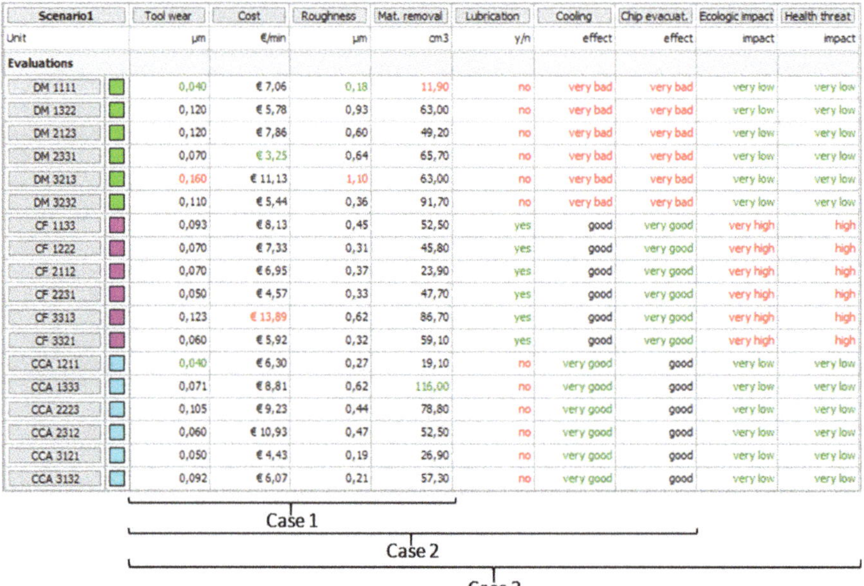

Figure 6. Input matrix with extended criteria set: 4 quantitative and 5 qualitative criteria.

The 18 alternatives, presented in Figure 6, used the results of experiments (Table 2) as their criteria evaluations. All four criteria were quantitative criteria. 18 alternatives were divided into 3 groups: 6 alternatives represented dry machining (DM) experiments, 6 alternatives represented conventional milling experiments with the application of cutting fluids (CF), and 6 alternatives represented milling experiments with the application of compressed cold air cooling (CCA).

Three case studies are used to discuss this issue of the selection of the optimal machining technique and parameter set (Figure 6). Case 1 used 4 quantitative criteria. Case 2 extended with 3 more qualitative criteria, which represented important technical side-effects: lubrication, cooling, and chip evacuation. Finally, Case 3 used the criteria of Case 2 plus additional ecologic and social criteria to fulfill sustainable machining criteria. Finally, an additional validation of Case 3 was made with criteria weights sensitivity analysis.

3.1. Case 1: Comparison of Milling Techniques Based on Quantitative Economic and Productivity Criteria

Case 1 represents a comparison of 18 different milling experiments based on 4 criteria. All criteria were quantitative, but two of them represented economic criteria (tool wear and cost) and two of them represented productivity criteria (roughness and material removal rate). Criteria evaluations of alternatives are given in Figure 6.

Figure 7 shows criteria parameters (weight, preference function type, and its parameters, etc.). Criteria weights can have a significant impact on the result, therefore equal criteria weights were given to both groups of criteria: economic criteria 50% and productivity criteria 50%, which results in 25% for each criterion (Figures 7 and 8).

Scenario1	Tool wear	Cost	Roughness	Mat. removal
Unit	µm	€/min	µm	cm3
Cluster/Group				
Preferences				
Min/Max	min	min	min	max
Weight	0,25	0,25	0,25	0,25
Preference Fn.	Linear	Linear	Linear	Linear
Thresholds	absolute	absolute	absolute	absolute
- Q: Indifference	0,005	€ 0,50	0,05	5,00
- P: Preference	0,050	€ 5,00	0,50	50,00

Figure 7. Criteria parameters and weights for Case 1.

The partial ranking by PROMETHEE I is presented in Figure 8a and shows that the first three ranked alternatives (CF 2231, CCA 3121, and CF 3321) are not completely comparable. They did not completely outrank each other, because if an alternative wants to outrank other alternatives it needs to have a higher positive flow (Phi+) and equal or lower negative flow (Phi-). However, when applying the net flow, i.e., when calculating PROMETHEE II ranking, alternative CF 2231 becomes the 1st ranked alternative (Figure 8b,c).

When analyzing the ranking, it became clear that this problem has some issues to solve. Namely, when looking at the first five ranked alternatives (Figure 8c): two of them were CF alternatives, one is a CCA alternative, and two are DM alternatives. The rest of the ranking shows the chaotic distribution of alternatives, as well. Therefore, it is clear that something is missing to produce a more stable comparison and ranking, and what is missing are some additional criteria that will enable proper comparison of machining techniques.

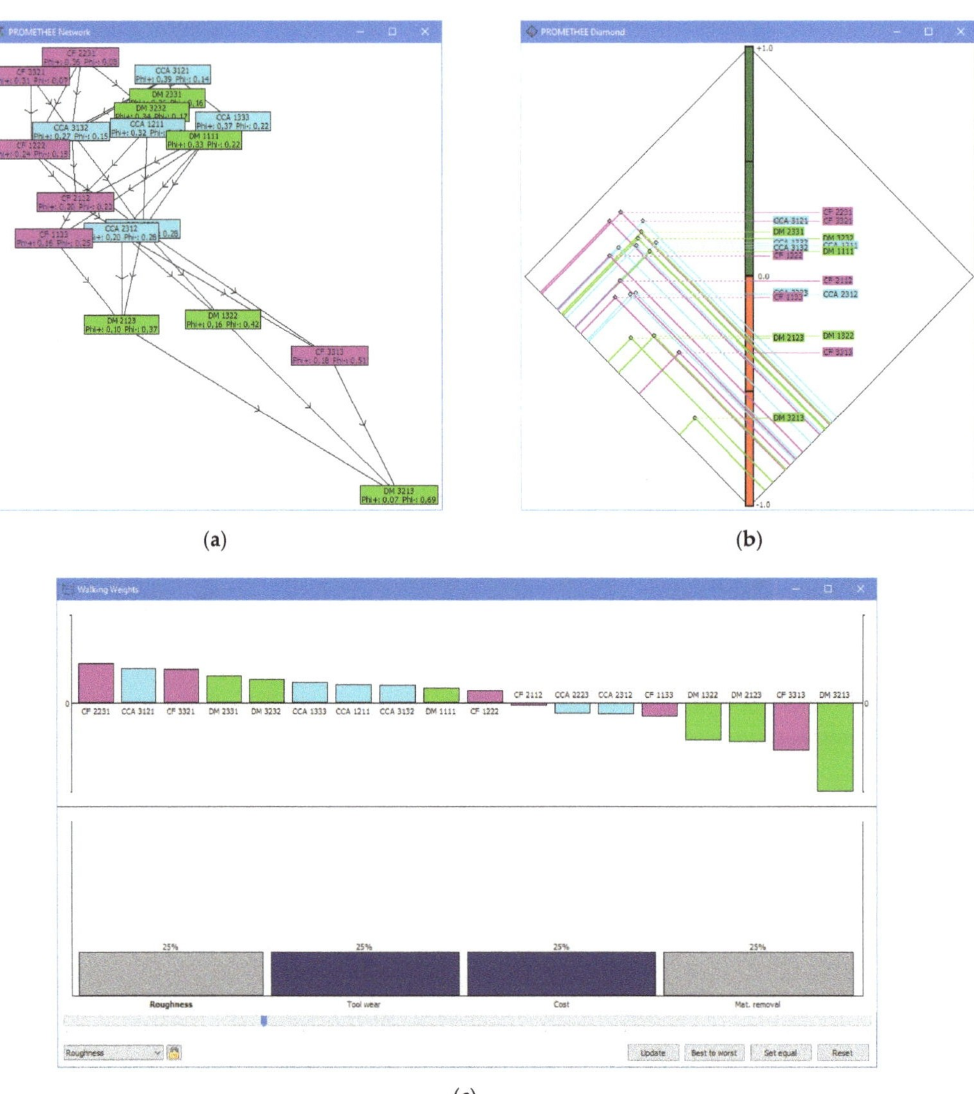

(a)

(b)

(c)

Figure 8. Case 1: (**a**) PROMETHEE I partial ranking of alternatives; (**b**) PROMETHEE II complete ranking of alternatives; (**c**) Distribution of criteria weights and ranking of alternatives (left to right: the best to the worst).

3.2. Case 2: Comparison of Milling Techniques Based on Quantitative and Qualitative Economic, and Productivity Criteria

Case 2 represents a comparison of 18 different milling experiments based on 7 criteria. The first four criteria were quantitative criteria, the same as in Case 1, but three additional qualitative criteria were added to the group of productivity criteria: lubrication, cooling and chip evacuation. These qualitative criteria better-described alternatives depending on the machining technique. Namely, lubrication had a good effect on the reduction of the tool wear during the machining process and was the only technique that provides lubrication in the application of cutting fluids (CF). Both application of cutting fluids (CF) and application of compressed cold air (CCA) provided cooling, but, CCA provided better cooling since the cold air was used, and cutting fluids were not cooled. Finally, chip evacuation was better with the application of cutting fluids (CF) than the application of compressed cold air (CCA), and dry machining (DM) had poor chip evacuation and cooling. It is clear that it is not easy to define proper quantitative descriptions of these criteria, so a qualitative description is used. The group of experts (mechanical engineers) were decision-makers that defined the qualitative evaluations of alternatives for these criteria. A 5-point Likert scale was used, so it was not hard to define these evaluations, but it is not so precise, as well. A more precise qualitative scale or quantitative description of these criteria will be the subject of future research.

Criteria evaluations of alternatives can be found in part of a matrix presented in Figure 6. Figure 9 shows criteria parameters (weight, preference function type, and its parameters, etc.), and, again, equal criteria weights were given to both groups of criteria: economic criteria 50% and productivity criteria 50%, which results with 25% for each of economic criteria and 10% for each of productivity criteria (Figures 9 and 10).

Scenario1	Tool wear	Cost	Roughness	Mat. removal	Lubrication	Cooling	Chip evacuat.
Unit	µm	€/min	µm	cm3	y/n	effect	effect
Cluster/Group	◆	◆	◇	◇	◇	◇	◇
Preferences							
Min/Max	min	min	min	max	max	max	max
Weight	0,25	0,25	0,10	0,10	0,10	0,10	0,10
Preference Fn.	Linear	Linear	Linear	Linear	Linear	Linear	Linear
Thresholds	absolute	absolute	absolute	absolute	absolute	absolute	absolute
- Q: Indifference	0,005	€ 0,50	0,05	5,00	0,00	0,00	0,00
- P: Preference	0,050	€ 5,00	0,50	50,00	1,00	4,00	4,00

Figure 9. Criteria parameters and weights for Case 2.

The partial ranking by PROMETHEE I is presented in Figure 10a, and shows that the 1st ranked alternative CF 2231 dominated above all other alternatives. The ranking by PROMETHEE II was more stable, since the DM alternatives were pushed to the bottom of the rank (Figure 10b,c). Now, the distribution of the first five ranked alternatives consisted of three CF alternatives and two CCA alternatives, and the best ranked DM alternative had 9th rank.

However, it is not clear from Figure 10 why this approach to the problem is more stable than the approach of Case 1. But, an additional look at GAIA (Geometrical Analysis for Interactive Aid) planes, which represent the projection of all criteria axes and alternatives on a single plane, can help to resolve this. Figure 11a represents the GAIA plot of the problem from Case 1, and it is completely chaotic with CF, CCA, and DM alternatives mutually mixed on the plane.

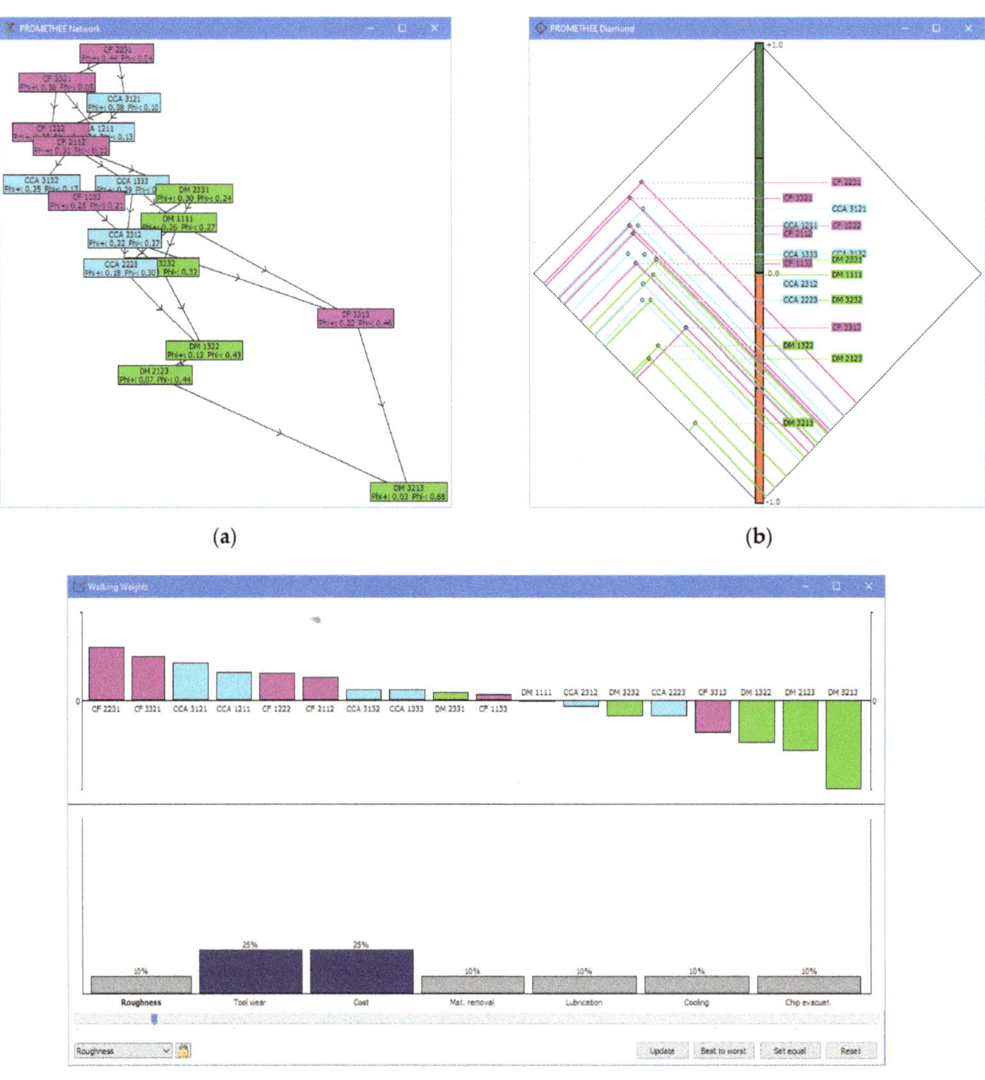

Figure 10. Case 2: (**a**) PROMETHEE I partial ranking of alternatives; (**b**) PROMETHEE II complete ranking of alternatives; (**c**) distribution of criteria weights and ranking of alternatives (left to right: the best to the worst).

The situation is different in Figure 11b, which represents the GAIA plot of the problem from Case 2. The GAIA plot of Case 2 shows three different clusters: CF alternatives cluster, CCA alternatives cluster, and DM alternatives cluster. So it can be concluded that Case 2 is a better approach to this problem than Case 1, however, Case 2 did not meet the sustainability criteria presented in Figure 4, therefore an additional case was needed.

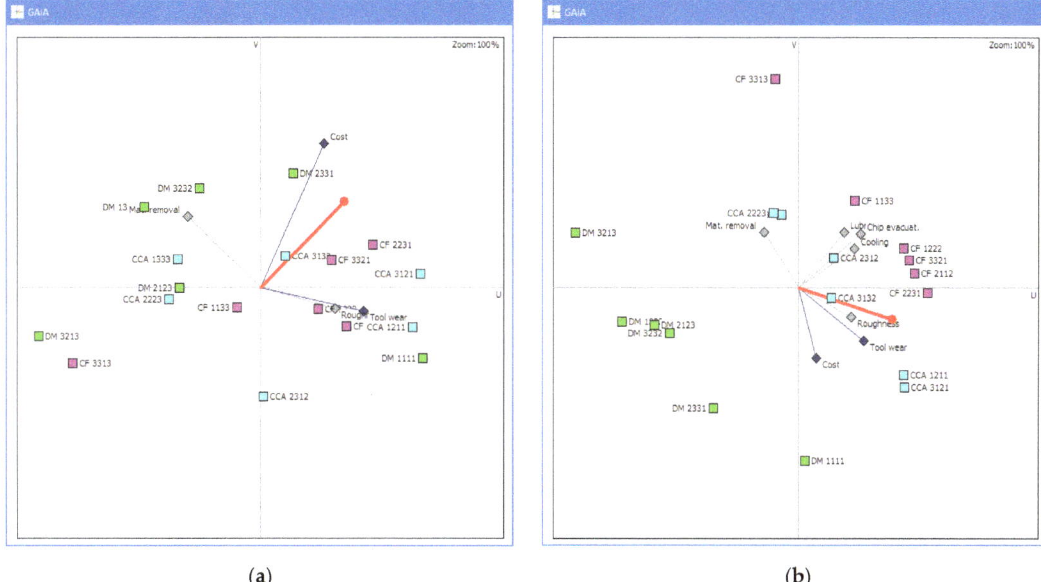

(a) (b)

Figure 11. Comparison of Case 1 and 2: (**a**) the GAIA plot of the problem from Case 1; (**b**) the GAIA plot of the problem from Case 2.

3.3. Case 3: Comparison of Milling Techniques Based on Sustainability: Economic, Productivity, Ecologic, and Social Criteria

Finally, Case 3 is a comparison of 18 different milling experiments based on sustainability criteria: economic (2 criteria), productivity (5 criteria), ecologic (1 criterion) and social (1 criterion). The first seven criteria were economic and productivity criteria, the same as in Case 2, but additionally ecologic (ecologic impact) and social (health threat) criteria were added. Ecologic impact represents a negative impact on the environment that CF alternatives have because of the cutting fluids that need to be processed and disposed of. Health threat represents the threat of the technique to human health and, again, CCA and DM do not represent a threat, but CF represents a serious threat because a worker's skin can be in contact with cutting fluids and there is a problem of cutting fluids' aerosol.

Criteria evaluations of alternatives are presented in Figure 6. Figure 12 shows the criteria parameters (weight, preference function type, and its parameters, etc.), and, again, equal criteria weights were given to all criteria groups: economic criteria 25%, productivity criteria 25%, ecologic criteria 25%, and social criteria 25%, which results in 12.5% for each of economic criteria, 5% for each of productivity criteria, 25% for ecologic criterion, and 25% for social criterion (Figures 12 and 13).

Scenario1	Tool wear	Cost	Roughness	Mat. removal	Lubrication	Cooling	Chip evacuat.	Ecologic impact	Health threat
Unit	μm	€/min	μm	cm3	y/n	effect	effect	impact	impact
Cluster/Group									
Preferences									
Min/Max	min	min	min	max	max	max	max	min	min
Weight	0,13	0,13	0,05	0,05	0,05	0,05	0,05	0,25	0,25
Preference Fn.	Linear	Linear	Linear	Linear	Linear	Linear	Linear	Linear	Linear
Thresholds	absolute	absolute	absolute	absolute	absolute	absolute	absolute	absolute	absolute
- Q: Indifference	0,005	€ 0,50	0,05	5,00	0,00	0,00	0,00	0,00	0,00
- P: Preference	0,050	€ 5,00	0,50	50,00	1,00	4,00	4,00	4,00	4,00

Figure 12. Criteria parameters and weights for Case 3.

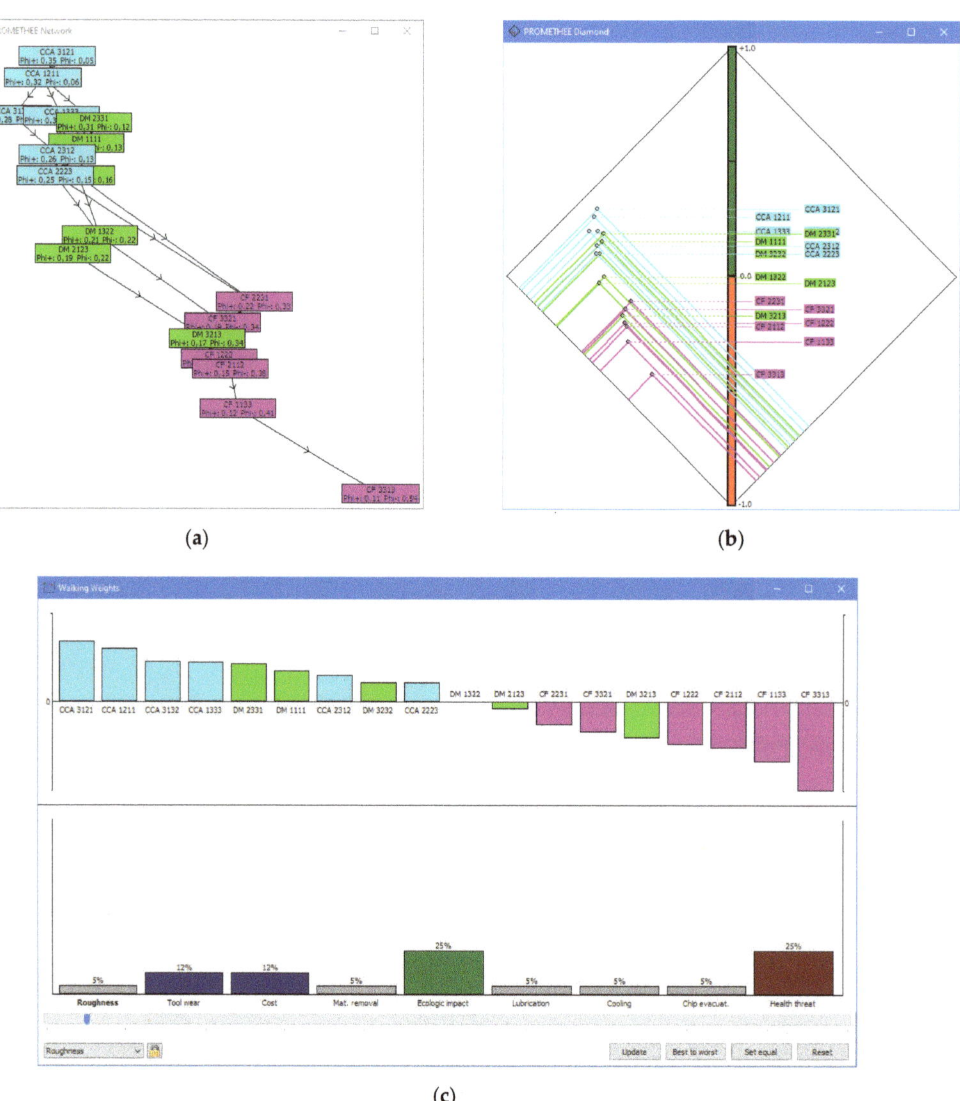

(a)

(b)

(c)

Figure 13. Case 3: (**a**) PROMETHEE I partial ranking of alternatives; (**b**) PROMETHEE II complete ranking of alternatives; (**c**) distribution of criteria weights and ranking of alternatives (left to right: the best to the worst).

The partial ranking by PROMETHEE I is presented in Figure 13a and shows that the 1st ranked alternative CCA 3121 dominates over the 2nd ranked alternative CCA 1211, and both of them dominate above all other alternatives. The ranking by PROMETHEE II shows complete domination of green technologies CCA and DM over CF technology (Figure 13b,c). Finally, the distribution of the first five ranked alternatives consists of the green technologies only: four CCA alternatives and one DM alternatives, and the best ranked CF alternative has 12th rank.

Furthermore, the GAIA plot indicates a stable problem and three different machining techniques are forming three different clusters of alternatives (Figure 14).

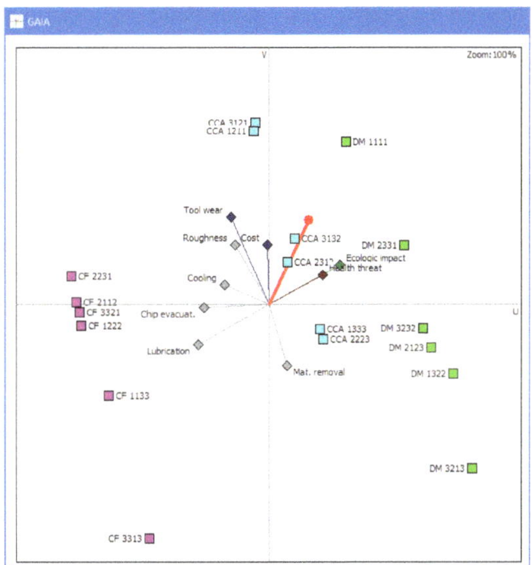

Figure 14. The GAIA plot of the problem from Case 3.

This case demonstrated the importance of using sustainability criteria that include ecologic and social criteria. Only in this case, the non-green cutting fluids machining technique did not get a high rank and would not be selected as a milling option.

3.4. Sustainability Criteria Weights Sensitivity Analysis

Before the final conclusion, some kind of validation of the sustainability-criteria-based approach (Case 3) must be made. A validation was made through criteria weights sensitivity analysis. An analysis was based on the determination of the criterion weight interval in which the rankings of the selected number of alternatives remained the same. An example of this analysis is given in Figure 15.

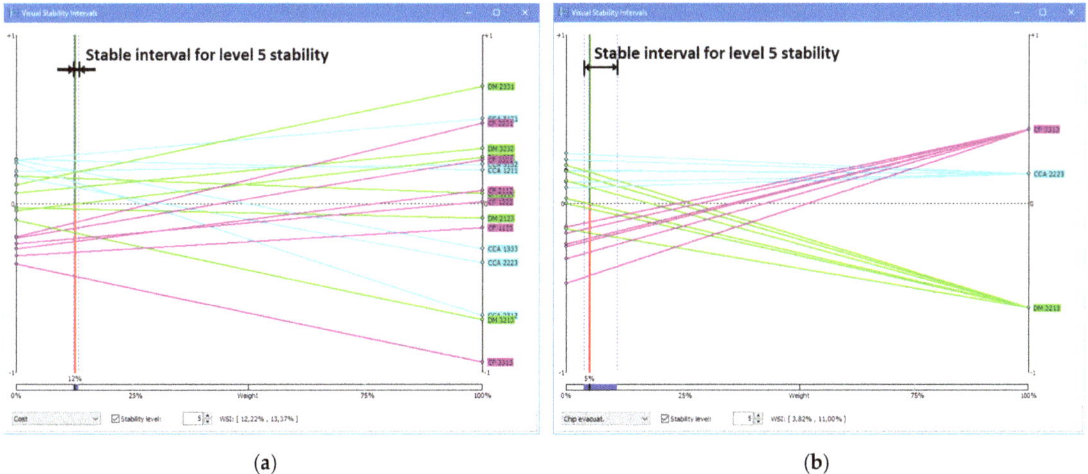

(a)   (b)

Figure 15. Comparison of stability intervals: (**a**) Stability interval (level 5) for criterion Cost; (**b**) Stability interval (level 5) for criterion Chip evacuation.

The level 5 stability was selected and it implied the weight interval in which the ranking of the first five alternatives was not changing. Other alternatives were perhaps swapping their ranks, but the ranking of the first five alternatives remains the same.

Figure 15a represents a stable interval for criterion Cost, and its weight is not so stable. Namely, criterion Cost had a weight of 12.5%, and it could have any weight between 12.22 and 13.37%, in order to keep the same order of the first five ranked alternatives. But, it was a very narrow interval compared to the interval of criterion Chip evacuation (Figure 15b), which was between 3.82 and 11%. Therefore, the weight interval of criterion Cost was actually unstable and the interval of criterion Chip evacuation was stable.

To validate this case, by checking the changes in ranking with the change of the criteria weights, a different set of criteria weights should be given. A common approach is to use equal weights, but not equal weights of criteria groups, but an equal weight of each criterion. In order to do so, all criteria weights were set to 11.1%. The results are given in Figure 16.

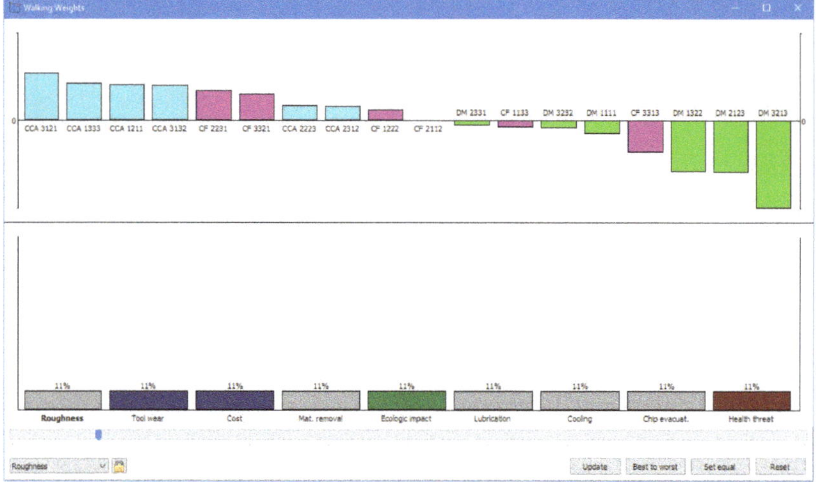

Figure 16. Distribution of criteria weights and ranking of alternatives for equal weights.

Figure 16 shows that, even in the case of equal weights, the green CCA technique with four best-ranked alternatives (CCA 3121, CCA 1333, CCA 1211, and CCA 3132) kept the lead over the non-green CF technique. Furthermore, CCA 3121 was the 1st ranked alternative in Case 3, as well. DM techniques, which were in the middle of the rank in Case 3, were at the bottom now. The reason is that DM alternatives have poor performance in terms of lubrication and cooling, which now had higher weights. To conclude, the validation of the sustainability-criteria-based approach has confirmed the approach, and also confirmed the advantages of the green CCA machining technique.

The criteria weights are the important issue and the analysis of different criteria sets, i.e., scenarios, definitely represent one of the most important aspects for future research. In this research, the same weight was used for each of four criteria groups: economic, productivity, ecologic, and social criteria. Of course, it would be interesting to see how different weights of these criteria groups affect the result. But criteria weights for the machining process cannot be defined randomly or approximately, so it represents a significant research issue, as presented by Kumar et al. [69]. The most common approaches are based on the usage of AHP method to mutually compare the importance of criteria and calculate their weights using the AHP procedure [70], or based on EWM (entropy weights method) that uses probability theory to compute uncertain information [71,72]. Both of the approaches will be considered for future research on this topic.

4. Conclusions

In this research, the decision support system was developed to support the selection of machining techniques and their parameters by taking the sustainability criteria into account. Three cases have been analyzed (Case 1, 2, and 3) to demonstrate the need and the importance of using sustainability criteria that include ecologic and social criteria. Case 1 and 2 used economic and productivity criteria only, and in both cases, the non-green cutting-fluids-based techniques got very high ranks. However, when ecologic and social criteria were added in Case 3, the green dry-machining-based and compressed-cold-air-based techniques achieved the highest ranks and suppressed cutting-fluids-based techniques towards the bottom. This kind of approach should be imperative: productivity cannot be maximized at the expense of the environment and human health. At the moment, and perhaps it will always remain that way, the problem is that the conventional cutting-fluids-based machining is more productive and more economical than green machining. So, the question is why should we invest in green machining techniques? Because it is investing in the future generations, so they could live in the same natural environment as present generations. However, a radical change in a mindset is needed to accept that fact, especially in managerial thinking. Nevertheless, the negative impact on the environment and human health can also be seen as an economical cost, although it is not easy to make such a calculation. If seen from that perspective, investing in green technologies reduces the economic cost of manufacturing. So, it is the change of the perspective that management and everybody else must accept. This research at least somehow contributes to that change. The main goal of further research will be the application of the developed approach and DSS to other machining and manufacturing processes, as well.

Author Contributions: Conceptualization, L.C. and M.M.; methodology, L.C. and M.M.; software, M.M.; validation, L.C., M.M., N.G. and M.C.Z.; writing—original draft preparation, L.C. and M.M.; writing—review and editing, L.C., M.M., N.G. and M.C.Z. All authors have read and agreed to the published version of the manuscript.

Funding: The publishing of this research was funded by the MZO-VIF INTEMON project.

Institutional Review Board Statement: Not applicable.

Informed Consent Statement: Not applicable.

Acknowledgments: Experimental data sets used in this research are from Luka Celent's doctoral dissertation, which was part of the MZOS 023-0692976-1742 research project.

Conflicts of Interest: The authors declare no conflict of interest.

References

1. King, N.; Keranen, L.; Gunter, K.; Sutherland, J. Wet Versus Dry Turning: A Comparison of Machining Costs, Product Quality, and Aerosol Formation. *SAE Tech. Pap.* **2001**, *2001*, 11. [CrossRef]
2. Kahle, L.R.; Gurel-Atay, E. *Communicating Sustainability for the Green Economy*, 1st ed.; Routledge: New York, NY, USA, 2014.
3. The United Nations. The 17 Sustainable Development Goals. Available online: https://sdgs.un.org/goals (accessed on 7 September 2021).
4. The Holy See, Encyclical Letter Laudato Si' of the Holy Father Francis on Care for Our Common Home. Available online: https://www.vatican.va/content/francesco/en/encyclicals/documents/papa-francesco_20150524_enciclica-laudato-si.html (accessed on 7 September 2021).
5. Khan, A.M.; Anwar, S.; Gupta, M.K.; AlFaify, A.; Hasnain, S.; Jamil, M.; Mia, M.; Pimenov, D.Y. Energy-Based Novel Quantifiable Sustainability Value Assessment Method for Machining Processes. *Energies* **2020**, *13*, 6144. [CrossRef]
6. Jamwal, A.; Agrawal, R.; Sharma, M.; Kumar, A.; Luthra, S.; Pongsakornrungsilp, S. Two decades of research trends and transformations in manufacturing sustainability: A systematic literature review and future research agenda. *Prod. Eng.* **2021**, 1–25. [CrossRef]
7. Byers, J.P. *Metalworking Fluids*, 3rd ed.; CRC Press: Boca Raton, FL, USA, 2017.
8. Glenn, T.; Van Antwerpen, F. Opportunities and Market Trend in Metalworking Fluids. *J. Soc. Tribol. Lubr. Eng.* **2004**, *54*, 31–34.
9. Abdala, H.S.; Baines, W.; McIntyre, G.; Slade, C. Development of novel sustainable neat-oil metal working fluids for stain-less steel and titanium alloy machining. Part 1. *Formul. Dev. Int. J. Adv. Manuf. Technol.* **2007**, *34*, 21–33. [CrossRef]

10. Debnath, S.; Reddy, M.M.; Yi, Q.S. Environmental friendly cutting fluids and cooling techniques in machining: A review. *J. Clean. Prod.* **2014**, *83*, 33–47. [CrossRef]

11. Young, P.; Byrne, G.; Cotterell, M. Manufacturing and the environment. *Int. J. Adv. Manuf. Technol.* **1997**, *13*, 488–493. [CrossRef]

12. Byrne, G.; Dornfeld, D.; Denkena, B. Advancing Cutting Technology. *CIRP Ann.* **2003**, *52*, 483–507. [CrossRef]

13. Soković, M.; Mijanović, K. Ecological aspects of the cutting fluids and its influence on quantifiable parameters of the cutting processes. *J. Mater. Process. Technol.* **2001**, *109*, 181–189. [CrossRef]

14. United States Environmental Protection Agency. National Occupational Exposure Survey (NOES). Available online: https://cfpub.epa.gov/si/si_public_record_Report.cfm?Lab=OEI&dirEntryID=132500 (accessed on 29 December 2021).

15. Ueno, S.; Shiomi, Y.; Yokota, K. Metalworking Fluid Hand Dermatitis. *Ind. Health* **2002**, *40*, 291–293. [CrossRef]

16. Mackerer, C.R. Health Effects of Oil Mists: A Brief Review. *Toxicol. Ind. Health* **1989**, *5*, 429–440. [CrossRef]

17. Thorne, P.S.; DeKoster, J.A.; Subramanian, P. Environmental Assessment of Aerosols, Bioaerosols, and Airborne Endotoxins in a Machining Plant. *Am. Ind. Hyg. Assoc. J.* **1996**, *57*, 1163–1167. [CrossRef]

18. Bartz, W.J. Lubricants and the environment. *Tribol. Int.* **1998**, *31*, 35–47. [CrossRef]

19. Davim, J.P. *Machining: Fundamentals and Recent Advances*, 2008th ed.; Springer: Berlin/Heidelberg, Germany, 2008.

20. Katna, R.; Suhaib, M.; Agrawal, N. Nonedible vegetable oil-based cutting fluids for machining processes—A review. *Mater. Manuf. Process.* **2020**, *35*, 1–32. [CrossRef]

21. Pusavec, F.; Kramar, D.; Krajnik, P.; Kopac, J. Transitioning to sustainable production—Part II: Evaluation of sustainable machining technologies. *J. Clean. Prod.* **2010**, *18*, 1211–1221. [CrossRef]

22. Kaynak, Y.; Gharibi, A. Progressive Tool Wear in Cryogenic Machining: The Effect of Liquid Nitrogen and Carbon Dioxide. *J. Manuf. Mater. Process.* **2018**, *2*, 31. [CrossRef]

23. Breque, M.; De Nul, L.; Petridis, A. *Industry 5.0—Towards a Sustainable, Human-Centric and Resilient European Industry*; Publications Office of the European Union: Luxembourg, 2021.

24. Müller, J. *Enabling Technologies for Industry 5.0—Results of a Workshop with Europe's Technology Leaders*; Publications Office of the European Union: Luxembourg, 2020.

25. Kagermann, H.; Wahlster, W.; Helbig, J. *Recommendations for Implementing the Strategic Initiative Industrie 4.0*; Heilmeyer und Sernau: Berlin, Germany, 2013.

26. Vacchi, M.; Siligardi, C.; Cedillo-González, E.I.; Ferrari, A.M.; Settembre-Blundo, D. Industry 4.0 and Smart Data as Enablers of the Circular Economy in Manufacturing: Product Re-Engineering with Circular Eco-Design. *Sustainability* **2021**, *13*, 10366. [CrossRef]

27. Enyoghasi, C.; Badurdeen, F. Industry 4.0 for sustainable manufacturing: Opportunities at the product, process, and system levels. *Resour. Conserv. Recycl.* **2021**, *166*, 105362. [CrossRef]

28. Gupta, K. A Review on Green Machining Techniques. *Procedia Manuf.* **2020**, *51*, 1730–1736. [CrossRef]

29. Sreejith, P.S.; Ngoi, B.K.A. Dry machining: Machining of the future. *J. Mater. Process. Technol.* **2000**, *101*, 287–291. [CrossRef]

30. Dudzinski, D.; Devillez, A.; Moufki, A.; Larrouquère, D.; Zerrouki, V.; Vigneau, J. A review of developments towards dry and high speed machining of Inconel 718 alloy. *Int. J. Mach. Tools Manuf.* **2004**, *44*, 439–456. [CrossRef]

31. Dinnie, K. Creating Corporate Reputations, Identity, Image, and Performance. *Eur. J. Mark.* **2003**, *37*, 114–147.

32. Naveed, M.; Arslan, A.; Javed, H.M.A.; Manzoor, T.; Quazi, M.M.; Imran, T.; Zulfattah, Z.M.; Khurram, M.; Fattah, I.M.R. State-of-the-Art and Future Perspectives of Environmentally Friendly Machining Using Biodegradable Cutting Fluids. *Energies* **2021**, *14*, 4816. [CrossRef]

33. Braga, D.U.; Diniz, A.E.; Miranda, G.W.A.; Coppini, N.L. Using a Minimum Quantity of Lubricant (MQL) and a Diamond Coated Tool in The Drilling of Aluminium-silicon Alloys. *J. Mater. Processing Technol.* **2002**, *122*, 127–138. [CrossRef]

34. Dixit, U.S.; Sarma, D.K.; Davim, J.P. *Environmentally Friendly Machining*, 2012th ed.; Springer: New York, NY, USA, 2012.

35. Su, Y.; He, N.; Li, A.; Iqbal, A.; Xiao, H.A.; Xu, S.; Qui, B.G. Refrigerated cooling air cutting of difficult-to-cut materials. *Inter-Natl. J. Mach. Tools Manuf.* **2007**, *47*, 927–933. [CrossRef]

36. Sharma, V.S.; Dogra, M.; Suri, N. Cooling techniques for improved productivity in turning. *Int. J. Mach. Tools Manuf.* **2009**, *49*, 435–453. [CrossRef]

37. Hilsch, R. The Use of the Expansion of Gases in a Centrifugal Field as Cooling Process. *Rev. Sci. Instrum.* **1947**, *18*, 108–113. [CrossRef]

38. Aronson, R.B. Vortex tube: Cooling with compressed air. *Mach. Des.* **1996**, *48*, 140–143.

39. Pinar, A.M.; Uluer, O.; Kirmaci, V. Optimization of counter flow Ranque—Hilsch vortex tube performance using Taguchi method. *Int. J. Refrig.* **2009**, *32*, 1487–1494. [CrossRef]

40. Rahman, M.; Kumar, A.S.; Salam, M.U.; Ling, M.S. Effect of Chilled Air on Machining Performance in End Milling. *Int. J. Adv. Manuf. Technol.* **2003**, *21*, 787–795. [CrossRef]

41. Yalçın, B.; Özgür, A.; Koru, M. The effects of various cooling strategies on surface roughness and tool wear during soft materials milling. *Mater. Des.* **2009**, *30*, 896–899. [CrossRef]

42. Timmerhaus, K.D.; Reed, R.P. *Cryogenic Engineering*, 1st ed.; Springer: New York, NY, USA, 2007.

43. Shokrani, A.; Dhokia, V.; Newman, S. Environmentally conscious machining of difficult-to-machine materials with regard to cutting fluids. *Int. J. Mach. Tools Manuf.* **2012**, *57*, 83–101. [CrossRef]

44. De Chiffre, L.; Andreasen, J.; Lagerberg, S.; Thesken, I.-B. Performance Testing of Cryogenic CO_2 as Cutting Fluid in Parting/Grooving and Threading Austenitic Stainless Steel. *CIRP Ann.* **2007**, *56*, 101–104. [CrossRef]
45. Kumar, K.K.; Choudhury, S. Investigation of tool wear and cutting force in cryogenic machining using design of experiments. *J. Mater. Process. Technol.* **2008**, *203*, 95–101. [CrossRef]
46. Pei, H.J.; Zheng, W.J.; Wang, G.C.; Wang, H.Q. Application of Biodegradable Cutting Fluids in High Speed Turning. *Adv. Mater. Res.* **2011**, *381*, 20–24. [CrossRef]
47. Celent, L. Implementation of Compressed Cold Air Using Vortex Tube in Milling Process [Implementacija Hlađenja Komprimiranim Hladnim Zrakom Korištenjem Vrtložne Cijevi U Postupku Glodanja]. Ph.D. Thesis, University of Split, Split, Croatia, 25 July 2014.
48. Bandaru, S.; Becerra, V.; Khanna, S.; Espargilliere, H.; Sevilla, L.T.; Radulovic, J.; Hutchinson, D.; Khusainov, R. A General Framework for Multi-Criteria Based Feasibility Studies for Solar Energy Projects: Application to a Real-World Solar Farm. *Energies* **2021**, *14*, 2204. [CrossRef]
49. Jamwal, A.; Agrawal, R.; Sharma, M.; Kumar, V. Review on multi-criteria decision analysis in sustainable manufacturing decision making. *Int. J. Sustain. Eng.* **2021**, *14*, 202–225. [CrossRef]
50. Watson, H.J. Revisiting Ralph Sprague's Framework for Developing Decision Support Systems. *Commun. Assoc. Inf. Syst.* **2018**, *42*, 363–385. [CrossRef]
51. Temuçin, T.; Tozan, H.; Valícek, J.; Harnicarova, M. A Fuzzy based decision support model for Non-traditional machining process selection. *Tech. Gaz.* **2013**, *20*, 787–793.
52. Taha, Z.; Rostam, S. A hybrid fuzzy AHP-PROMETHEE decision support system for machine tool selection in flexible manufacturing cell. *J. Intell. Manuf.* **2012**, *23*, 2137–2149. [CrossRef]
53. Alberti, M.; Ciurana, J.; Rodríguez, C.A.; Özel, T. Design of a decision support system for machine tool selection based on machine characteristics and performance tests. *J. Intell. Manuf.* **2011**, *22*, 263–277. [CrossRef]
54. Balazinski, M.; Bellerose, M.; Czogala, E. Application of fuzzy logic techniques to the selection of cutting parameters in machining processes. *Fuzzy Sets Syst.* **1994**, *63*, 307–317. [CrossRef]
55. Niamat, M.; Sarfraz, S.; Ahmad, W.; Shehab, E.; Salonitis, K. Parametric Modelling and Multi-Objective Optimization of Electro Discharge Machining Process Parameters for Sustainable Production. *Energies* **2020**, *13*, 38. [CrossRef]
56. Ming, W.; Shen, F.; Zhang, G.; Liu, G.; Du, J.; Chen, Z. Green machining: A framework for optimization of cutting parameters to minimize energy consumption and exhaust emissions during electrical discharge machining of Al 6061 and SKD 11. *J. Clean. Prod.* **2021**, *285*, 124889. [CrossRef]
57. Wittbrodt, P.; Paszek, A. Decision support system of machining process based on the elements of fuzzy logic. *Int. J. Mod. Manuf. Technol.* **2015**, *7*, 81–85.
58. Vidal, A.; Alberti, M.; Ciurana, J.; Casadesús, M. A decision support system for optimising the selection of parameters when planning milling operations. *Int. J. Mach. Tools Manuf.* **2005**, *45*, 201–210. [CrossRef]
59. Plaza, M.; Zębala, W.; Matras, A. Decision system supporting optimization of machining strategy. *Comput. Ind. Eng.* **2019**, *127*, 21–38. [CrossRef]
60. Shin, S.-J.; Kim, D.B.; Shao, G.; Brodsky, A.; Lechevalier, D. Developing a decision support system for improving sustainability performance of manufacturing processes. *J. Intell. Manuf.* **2015**, *28*, 1421–1440. [CrossRef]
61. Khan, A.M.; Jamil, M.; Salonitis, K.; Sarfraz, S.; Zhao, W.; He, N.; Mia, M.; Zhao, G. Multi-Objective Optimization of Energy Consumption and Surface Quality in Nanofluid SQCL Assisted Face Milling. *Energies* **2019**, *12*, 710. [CrossRef]
62. Ransikarbum, K.; Pitakaso, R.; Kim, N.; Ma, J. Multicriteria decision analysis framework for part orientation analysis in additive manufacturing. *J. Comput. Des. Eng.* **2021**, *8*, 1141–1157. [CrossRef]
63. Ransikarbum, K.; Khamhong, P. Integrated Fuzzy Analytic Hierarchy Process and Technique for Order of Preference by Similarity to Ideal Solution for Additive Manufacturing Printer Selection. *J. Mater. Eng. Perform.* **2021**, *9*, 1–12. [CrossRef]
64. Boumaiza, A.; Sanfilippo, A.; Mohandes, N. Modeling multi-criteria decision analysis in residential PV adoption. *Energy Strat. Rev.* **2021**, *39*, 100789. [CrossRef]
65. Chanthakhot, W.; Ransikarbum, K. Integrated IEW-TOPSIS and Fire Dynamics Simulation for Agent-Based Evacuation Modeling in Industrial Safety. *Safety* **2021**, *7*, 47. [CrossRef]
66. Mladineo, M.; Veža, I. Ranking enterprises in terms of competences inside regional production network. *Croat. Oper. Res. Rev.* **2013**, *4*, 65–75.
67. Nemery, P. On the Use of Multicriteria Ranking Methods in Sorting Problems. Ph.D. Thesis, Université Libre de Bruxelles, Brussels, Belgium, November 2008.
68. Brans, J.P.; Vincke, P.; Mareschal, B. How to select and how to rank projects: The Promethee method. *Eur. J. Oper. Res.* **1986**, *24*, 228–238. [CrossRef]
69. Kumar, R.; Bilga, P.S.; Singh, S. Multi-objective optimization using different methods of assigning weights to energy consumption responses, surface roughness and material removal rate during rough turning operation. *J. Clean. Prod.* **2017**, *164*, 45–57. [CrossRef]
70. Jajac, N.; Knezic, S.; Marovic, I. Decision support system to urban infrastructure maintenance management. *Organ. Technol. Manag. Constr.* **2009**, *1*, 72–79.

71. Sidhu, A.S.; Singh, S.; Kumar, R.; Pimenov, D.Y.; Giasin, K. Prioritizing Energy-Intensive Machining Operations and Gauging the Influence of Electric Parameters: An Industrial Case Study. *Energies* **2021**, *14*, 4761. [CrossRef]
72. Kumar, R.; Singh, S.; Bilga, P.S.; Jatin, K.; Singh, J.; Singh, S.; Scutaru, M.-L.; Pruncu, C.I. Revealing the benefits of entropy weights method for multi-objective optimization in machining operations: A critical review. *J. Mater. Res. Technol.* **2021**, *10*, 1471–1492. [CrossRef]

Article

Integrated Functional Safety and Cybersecurity Evaluation in a Framework for Business Continuity Management

Kazimierz T. Kosmowski [1], Emilian Piesik [1,*], Jan Piesik [2] and Marcin Śliwiński [1]

[1] Faculty of Electrical and Control Engineering, Gdansk University of Technology, G. Narutowicza 11/12, 80-233 Gdansk, Poland; kazimierz.kosmowski@pg.edu.pl (K.T.K.); marcin.sliwinski@pg.edu.pl (M.Ś.)

[2] Michelin Polska Sp. z o.o., St. W. Leonharda 9, 10-454 Olsztyn, Poland; jan.piesik@michelin.com

* Correspondence: emilian.piesik@pg.edu.pl

Abstract: This article outlines an integrated functional safety and cybersecurity evaluation approach within a framework for business continuity management (BCM) in energy companies, including those using Industry 4.0 business and technical solutions. In such companies, information and communication technology (ICT), and industrial automation and control system (IACS) play important roles. Using advanced technologies in modern manufacturing systems and process plants can, however, create management impediments due to the openness of these technologies to external systems and networks via various communication channels. This makes company assets and resources potentially vulnerable to risks, e.g., due to cyber-attacks. In the BCM-oriented approach proposed here, both preventive and recovery activities are considered in light of engineering best practices and selected international standards, reports, and domain publications.

Keywords: functional safety; cybersecurity; BCM; Industry 4.0; information technology; industrial control system

Citation: Kosmowski, K.T.; Piesik, E.; Piesik, J.; Śliwiński, M. Integrated Functional Safety and Cybersecurity Evaluation in a Framework for Business Continuity Management. *Energies* **2022**, *15*, 3610. https://doi.org/10.3390/en15103610

Academic Editor: Marko Mladineo

Received: 17 March 2022
Accepted: 12 May 2022
Published: 15 May 2022

Publisher's Note: MDPI stays neutral with regard to jurisdictional claims in published maps and institutional affiliations.

1. Introduction

Industrial companies nowadays, including those implementing Industry 4.0 smart technologies, face potential safety and security problems due to their use of open systems and networks for communication and control [1–3]. The same concerns exist with respect to the energy systems within critical infrastructure such as power plants for producing electricity and/or heat from various energy sources, including coal, oil, natural gas, biogas, and renewable energy sources.

In order for power plants and distributed industrial systems to be economically effective, they should be reliable in continuous operation mode, or with the highest achievable availability when their operation is required on demand (e.g., during peak load of the electrical grid or during an abnormal state due to dependent or cascade failure leading to emergency conditions). These issues can be considered from a business continuity management (BCM) [4,5] point of view.

A traditional RAMS&S (reliability, availability, maintainability, safety, and security) methodology [6,7] can support elements of BCM in the life cycle, however, it insufficient due to its need to consider various impact factors, including the human and organizational factors. Certain aspects of BCM can be analyzed regarding performability engineering, as analysed and emphasised by Misra [8]. An interdisciplinary review of business continuity issues from the perspective of information systems, directed towards proposing an integrated framework, was published by Niemimaa [9]. These issues are lately of increasing attention to insurance companies [10,11].

Relatively new aspects in BCM analysis are connected to information and communication technology (ICT) and industrial control systems (ICS) that operate within computer systems and networks using wired or wireless communication channels. These systems and networks have been considered in several publications and research reports from the perspective of systems engineering [12–15] and cyber-physical systems [16,17]. Several research projects have been undertaken concerning the integrated analysis of ICS safety and security [18,19]. Interesting research works have been published concerning business continuity management, for instance an article [20] and monograph [21]. The functional safety and cybersecurity issues of industrial automation and control systems (IACS) have lately been emphasized as especially important in the design and operation of hazardous industrial plants and critical infrastructure systems [22–25].

Several security-related issues of the industrial automation and control system (IACS) have been considered in the context of protection solutions proposed for improving IACS security as proposed in the IEC 62443 standard [26]. The dependability and safety integrity of the safety-related part of the ICS are discussed with regard to the generic functional safety standard IEC 61508 in [27].

The remainder of this article is structured as follows. Section 2 provides a basic overview of functional safety and cybersecurity aspects related to business continuity management and the basic requirements in the context of risk evaluation within the life cycle; in addition, a BCM framework is proposed for business continuity planning in industrial companies. Section 3 outlines an integrated dependability, safety, and security management framework for industrial companies, including BCM aspects. In Section 4, a case study is presented to demonstrate the application of the proposed integrated approach. In the conclusions, the significance of adequately treating ICT and IACS within BCM activities in Industry 4.0 is emphasized.

2. Brief Presentation of the Framework and Components

2.1. Overview of IT and OT Systems and Their Convergence

The convergence of information technology (IT) and operational technology (OT) creates both new opportunities and new challenges. The data flows outside and into plant networks inevitably lead to additional threats and increased security-related risks. One of the biggest challenges facing the industrial sector is understanding the risks involved in potential cyberattacks, which are already being observed; these risks can emerge when companies adopt Industry 4.0 technologies, including Industrial Internet of Things (IIoT) technologies and tools. The management staff of industrial companies are becoming more aware about the magnitude of the gap between the priorities recognized by teams responsible for operational technology (OT) and those recognized by information technology (IT) professionals. This gap often impacts new cybersecurity initiatives.

In order to explain the issues involved, it is necessary to begin with a model industrial system. The traditional reference model is based on the ISA99 series of standards derived from the generic model of ANSI/ISA-95.00.01 (Enterprise-Control System Integration), and represents the manufacturing system using five functional and logical levels (Figure 1). These levels are often assigned to two classes, namely, Operational Technology (OT) and Information Technology (IT), with their own relevant security zones. The zero level defines the actual physical processes. The first level of activities involved in sensing and manipulating physical processes include intelligent devices. The second level includes control systems (e.g., Programmable Logic Controllers). The third level, site manufacturing and control, includes an ICS/SCADA system with a relevant Human–System Interface (HSI) and the Manufacturing Execution System (MES). The fourth level, enterprise business planning and logistics, comprises an Enterprise Resource Planning (ERP) system for effectively management and coordination of the business and enterprise resources required for manufacturing processes. Finally, the fifth level is the enterprise network for business and logistics activities, which can now be supported using applications based on Cloud Technology (CT) [28,29].

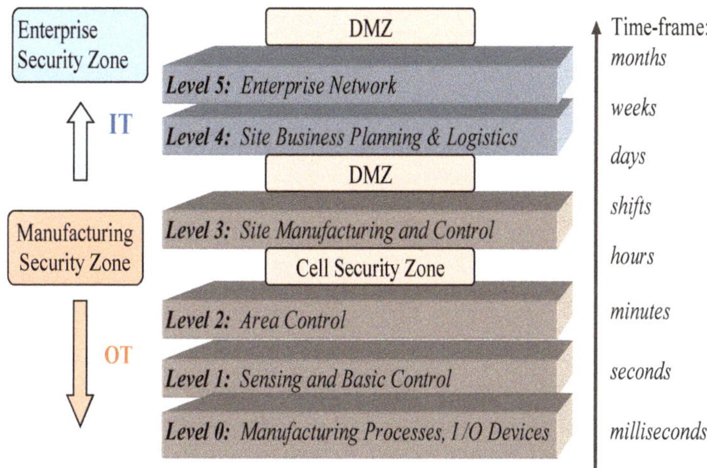

Figure 1. Traditional reference model of an industrial system based on the ANSI/ISA95 standard.

In an open manufacturing system, assigning safety and security-related requirements requires the special attention of designers and operators [3,30,31].

From an information security point of view, an important requirement and solution is to prioritize segmentation of the complex industrial computer system and network, distinguishing cell security zones and designing a Demilitarized Zone (DMZ), as illustrated in Figure 1.

The DMZ is sometimes referred to as a perimeter network or screened subnet, and is a physical or logical subnetwork for controlling and securing internal data and services from an organization's external services using an untrusted (usually larger) network such as a corporate-wide area network (WAN), the Internet, or a cloud technology (CT).

Thus, the purpose of a DMZ is to add a layer of security to an organization's local area network (LAN); an external network node can access only what is exposed in the DMZ, while the rest of the organization's network is firewalled [1,30].

An actual list of internal and external influences, hazards, and threats should be considered during the design and operation of the OT and IT systems and networks. Basic features of these systems are presented in Figure 2.

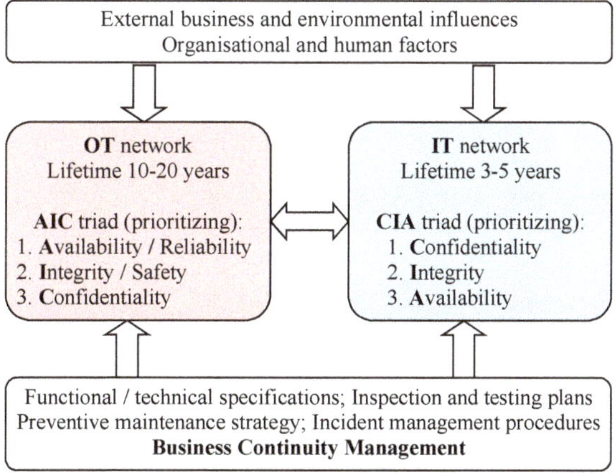

Figure 2. Basic features characterizing OT and IT systems and networks [23].

While the expected lifetime of OT systems is typically evaluated in the range of 10–20 years, this drops to only 3–5 years in the case of IT systems [23]. In characterizing the OT system, the AIC triad (Availability, Integrity, and Confidentiality) is often used to prioritize basic requirements, while the CIA triad (Confidentiality, Integrity, and Availability) is used to characterize the IT network.

The safety and security of both OT and IT systems and networks are dependent on various external and internal influences, including organizational and human factors [32]. Traditionally, a general MTE (Man-Technology-Environment) approach has been proposed for systemic analyses and management in the life cycle of industrial installations. An interesting framework for dealing with complex technical systems is offered by systems engineering (SE) [13]. The industrial automation and control system (IACS) [26,33] can be considered as a cyber-physical system [17,34,35].

2.2. Functional Safety of OT Systems

For high dependability and safety of the OT system, an operational strategy within BCM should be elaborated that includes inspections and periodical testing of safety-related control systems, for instance, electrical/electronic/programmable electronic (E/E/PE) systems [27] and safety instrumented systems (SIS) [36], including their sensors and the equipment under control (EUC).

The operational equipment of manufacturing lines (machinery, drives, operational control systems, etc.) requires an advanced preventive maintenance strategy to be implemented in order to achieve the required high OT availability and reduce the risk of outages and related production losses. Incident management procedures must be developed to reduce the risk of potentially hazardous events leading to major losses.

A set of safety functions are implemented in the safety-related ICS of required safety integrity levels (SILr), determined in the risk assessment process in relation to the criteria defined [12], to be assigned, for instance, to the E/E/PE or SIS systems (see the OT block in Figure 3).

Two different requirements must be specified to ensure an appropriate level of functional safety [37]:

- The requirements imposed on the performance of safety functions designed for hazard identification;
- The safety integrity requirements, i.e., the probability that a safety function will be performed in a satisfactory way when a potentially hazardous situation occurs.

Safety integrity is defined as the probability that a safety-related system, such as the E/E/PE system or SIS, will satisfactorily perform a defined safety function under all stated conditions within a given time. For safety-related ICS in which a defined safety function is implemented two probabilistic criteria must be defined, as presented in Table 1 for four categories of the SIL [27], namely:

- The probability of failure on demand average (PFD_{avg}) of the safety-related ICS in which the considered safety function is implemented, operating in a low-demand mode (LDM); or
- The probability of dangerous failure per hour (PFH) of the safety-related ICS operating in a high- or continuous-demand mode (HCM).

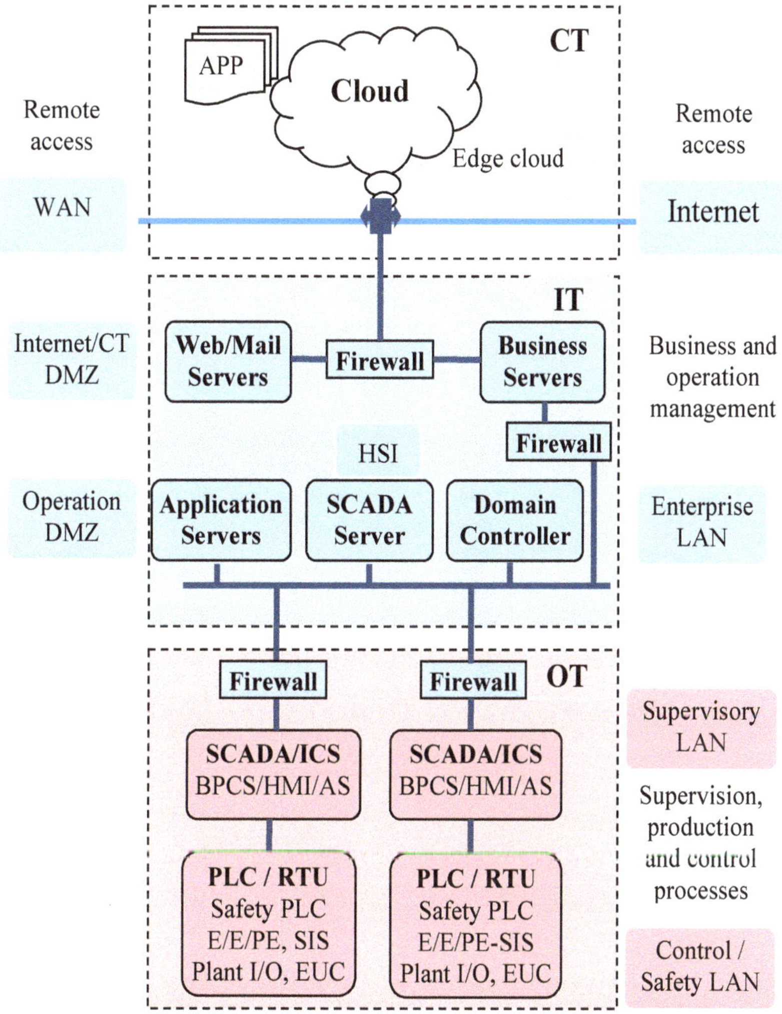

Figure 3. Typical ICT and ICS architecture including OT, IT, and CT.

Table 1. Categories of SIL and probabilistic criteria to be assigned to safety-related ICS operating in LDM or HCM.

SIL	PFD_{avg}	PFH [h^{-1}]
4	$[10^{-5}, 10^{-4})$	$[10^{-9}, 10^{-8})$
3	$[10^{-4}, 10^{-3})$	$[10^{-8}, 10^{-7})$
2	$[10^{-3}, 10^{-2})$	$[10^{-7}, 10^{-6})$
1	$[10^{-2}, 10^{-1})$	$[10^{-6}, 10^{-5})$

The SIL requirements assigned for the safety-related ICS to be designed for implementing a specified safety function stem from the results of the risk analysis and evaluation meant to reduce the risk of losses by sufficiently considering specified risk criteria, namely, for individual risk and/or group or societal risk [27].

If societal risk is of interest, analyses can generally be oriented on three distinguished categories of loss, namely [27,36,38], health (*H*), environment (*E*), and material (*M*) damage; then, the SIL required (SIL$_r$) for a particular safety function is determined as follows:

$$\text{SIL}_r = \max\left(\text{SIL}_r^H,\ \text{SIL}_r^E,\ \text{SIL}_r^M\right) \tag{1}$$

As mentioned above, SIL verification can generally be carried out for either of two operation modes, namely, LDM or HCM. The former is characteristic of the process industry [36], while the latter is typical for machinery [39], railway transportation systems, and the monitoring and real-time control of any installation using an ICS/SCADA system.

Management of the OT system and IACS, including safety-related lifecycle ICS, can be challenging; in industrial practice, it is difficult to achieve the above-specified requirements concerning the AIC triad (see Figure 3) for various reasons. Nevertheless, these systems contribute significantly to the realization of required quality and quantity of products in time, and influence overall equipment effectiveness (OEE). No less important are the functional safety and cybersecurity issues regarding the requirements and criteria discussed above.

The following items should be specified for implementation in industrial practice:

- A plan for operating and maintaining E/E/PE safety-related systems or SIS;
- Operation, maintenance, and repair procedures for these systems over their whole life cycle;

Implementation of these items must include initiation of the following actions:

- Implementing procedures;
- Following maintenance schedules;
- Maintaining relevant documentation;
- Periodically carrying out functional safety audits;
- Documenting any modifications to the hardware and software in E/E/PE systems.

Thus, all modifications that have an impact on the functional safety of any E/E/PE safety-related system must initiate a return to an appropriate phase of the overall E/E/PE system or software safety lifecycles. All subsequent phases must then be carried out in accordance with the procedures specified for the specific phases regarding the requirements in the above-mentioned standards.

For each phase of the overall functional safety lifecycles, a plan for verification and validation should be established concurrently with the development of consecutive phases. The verification plan must document or refer to the criteria, techniques, and tools to be used in verification activities.

Chronological documentation of operation, repair, and maintenance of safety-related systems should be maintained and must include the following information:

- The results of functional safety audits and tests;
- Documentation on the time and cause of demands on E/E/PE safety-related systems in actual operation the performance of the E/E/PE safety-related systems when subject to those demands, and any faults found during routine testing and maintenance;
- Documentation of any modifications made to safety-related ICS, including equipment under control (EUC).

The requirements concerning chronological documentation should be sufficiently detailed for the specific context of safety-related ICS operations [27,36,39].

2.3. Cybersecurity of IT Systems

From a cybersecurity perspective, the systems and networks used within the business environment (level 4 of the ISA95 model in Figure 1) should be considered as potentially insecure, as they contain complex interdependent hardware and software (see simplified architecture in Figure 3), remote access paths, and external communications. Therefore, IT and OT systems with the access to the Internet and/or a wide area network (WAN), or when the cloud technology (CT) is used, should be secured at the required security assurance level (SAL) for assignment to respective zones [26]. It has been postulated that the SAL assigned to the relevant domain should be included when verifying the safety integrity level (SIL) of safety-related ICS in which a specified safety function is to be implemented [12,40].

Security-related risks can be mitigated through the combined efforts of component suppliers, the machinery manufacturer, the system integrator, and the machinery final end user (with the company owner responsible) [26,33]. Generally, the response to a security risks should be as follows [41]:

(a)   Eliminate the security risk by design (avoiding vulnerabilities);
(b)   Mitigate the security risk by risk reduction measures (limiting vulnerabilities);
(c)   Provide information about residual security risks and measures to be adopted by the user.

The IEC 62443 standard [26] proposes an approach to dealing systematically with security-related issues in IACS. Four security levels (SLs) have been defined, understood as a confidence measure for ensuring that the IACS is free from vulnerabilities and will function in the intended manner. These SLs are suggested in the standard IEC 63074 [41] for dealing with the security of safety-related ICS designed for the operation of manufacturing plants.

These levels (numbered from 1 to 4, see Table 2) represent a piece of qualitative information addressing the relevant protection scope of the domain or zone considered in the evaluation against potential violations during safety-related ICS operation in a zone.

Table 2. Security levels and protection description of the IACS domain [26,41].

Security Levels	Description
SL 1	Protection against casual or coincidental violation
SL 2	Protection against intentional violation using simple means with low resources, generic skills, and low motivation
SL 3	Protection against intentional violation using sophisticated means with moderate resources, IACS-specific skills, and moderate motivation
SL 4	Protection against intentional violation using sophisticated means with extended resources, IACS-specific skills, and high motivation

The relevant SL number from 1 to 4 should be assigned to seven consecutive foundational requirements (FRs) relevant within the domain considered [26]:

FR 1—Identification and authentication control (IAC);
FR 2—Use control (UC);
FR 3—System integrity (SI);
FR 4—Data confidentiality (DC);
FR 5—Restricted data flow (RDF);
FR 6—Timely response to events (TRE);
FR 7—Resource availability (RA).

Thus, it is suggested that dependability and security-related evaluations apply a defined vector of relevant FRs from those specified above. Such a vector might be defined for the security-related requirements for a zone, conduit, component, or system. It contains the general integer numbers characterizing the SL from 1 to 4 (or 0 if not relevant) to be assigned to consecutive FR.

A general format of the security assurance level (SAL) to be evaluated for a given domain is defined as a function of [FRs] [26]:

$$\text{SAL} \times ([\text{FRs}]\ \text{domain}) = f\,[\text{IAC UC SI DC RDF TRE RA}] \tag{2}$$

2.4. Integrated Functional Safety and Cybersecurity Evaluation

Assigning the SAL to the domain or zone as an integer number from 1 to 4 [37,42] can present problems. To overcome this difficulty, the security indicator SI^{Do} for a domain (Do) can be defined [40] to determine security levels SL_i for a set (Re) of relevant fundamental requirements (FR$_i$) with relevant weights w_i evaluated based on the opinions of ICT and ICS experts. This indicator is a real number from the interval (1.0, 4.0) calculated using the following formula:

$$SI^{Do} = \sum_{i \in \text{Re}} w_i SL_i, \quad \sum_i w_i = 1 \tag{3}$$

Four intervals of the domain security index SI^{Do} (from SI^{Do1} to SI^{Do4}) are proposed in the first column of Table 3 for assigning an SAL category integer number from 1 to 4. This approach corresponds with that used in earlier publications for attributing an SAL to the domain based on the dominant SL_i for the relevant fundamental requirements, FR$_i$.

Table 3. Proposed correlation between SI^{Do}/SAL for the evaluated domain and final SIL to be attributed to the safety-related ICS of a critical installation.

Security Indicator SI^{Do}/SAL	SIL Verified According to IEC 61508 *			
	1	2	3	4
SI^{Do1}∈[1.0, 1.5)/SAL 1	SIL 1	SIL 1	SIL 1	SIL 1
SI^{Do2}∈[1.5, 2.5)/SAL 2	SIL 1	SIL 2	SIL 2	SIL 2
SI^{Do3}∈[2.5, 3.5)/SAL 3	SIL 1	SIL 2	SIL 3	SIL 3
SI^{Do4}∈[3.5, 4.0]/SAL 4	SIL 1	SIL 2	SIL 3	SIL 4

* verification includes the architectural constraints regarding S_{FF} and HFT of subsystems.

Three types of vectors describing SL_i for consecutive FR$_i$ of a domain can be distinguished [24]:

- SL-T (target SAL)—Desired level of security;
- SL-C (capability SAL)—Security level that the device can provide when properly configured;
- SL-A (achieved SAL)—Actual level of security of a particular device.

Proposed correlations between the security index to be assigned to the domain SI^{Do}/SAL and the final SIL attributed to the safety-related ICS in a hazardous installation are presented in Table 3. It was assumed that SILs were verified according to IEC 61508 requirements based on the results of probabilistic modelling [12,43], regarding potential common cause failures (CCFs) and the influence of the human and organizational factors regarding architectural constraints for the evaluated S_{FF} and HFT of the E/E/PE subsystems (see explanations above). Thus, SIL verification requires probabilistic modelling of the safety-related ICS of the proposed architecture regarding the S_{FF} and HFT of the subsystems.

2.5. Scope of BCM

Business continuity management (BCM) is usually understood as the capability and specified activity of an organization to continue delivery of products and/or services of required quality within acceptable time frames at a predefined capacity relating to the scale of potential disruptions [4].

A disruption is defined as an incident, whether anticipated or unanticipated, that causes an unplanned negative deviation from the expected delivery of products and services according to an organization's objectives. An objective is the result to be achieved. The objective can be strategic, tactical, or operational.

The objective can be expressed in other ways, e.g., as an intended outcome, a purpose, an operational criterion, or using other words with similar meaning (e.g., aim, goal, or target). Objectives can relate to different disciplines (such as financial, health and safety, and environmental objectives) and can apply at different levels (such as strategic, organization-wide, project, product, and process).

The BCM can be considered an integral part of a holistic risk management that safeguards the interests of the organization's key stakeholders, reputation, brand, and value by creating activities through [10]:

- Identifying potential threats that might cause adverse impacts on an organization's business operations, and associated risks;
- Providing a framework for building resilience for business operations;
- Providing capabilities, facilities, processes, and elaborated action task lists, etc., for effective responses to disasters and failures.

An event can be an occurrence or change in a particular set of circumstances that could have several causes and several consequences. An abnormal event due to a hazard or threat is considered a risk source. An emergency is a result of a sudden, urgent, usually unexpected occurrence or event requiring immediate action. It is a disruption or condition that can be anticipated or prepared for, although seldom exactly foreseen [44–46].

The organization must implement and maintain a systematic risk assessment process. Such a process could be carried out, for instance, in accordance with the ISO 31000 standard. As shown in Figure 4, an organization should:

(a)   Identify risks of disruption to the organization's prioritized activities and their supporting resources;
(b)   Systematically analyze and assess risks of disruption;
(c)   Evaluate risks of disruptions that require adequate treatment.

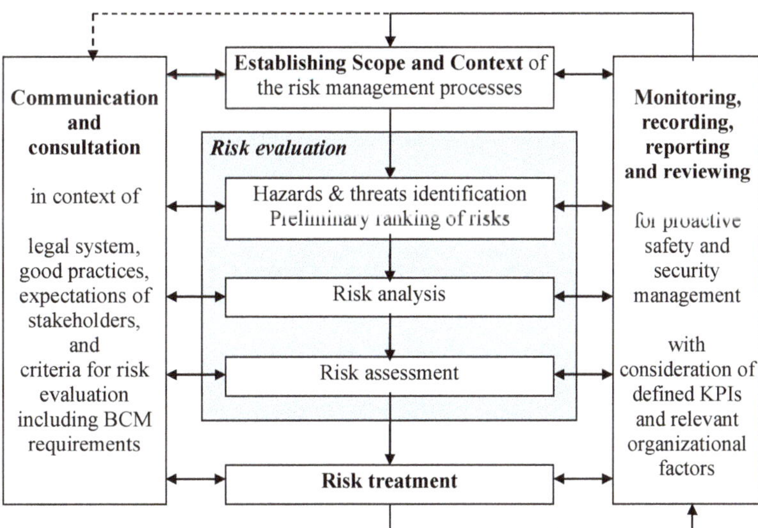

Figure 4. Risk management process (based on [47]).

Risk evaluation is considered an overall process of hazard/threat identification, risk analysis, and risk assessment [28,47]. Risk management is a process of coordinating activities in order to direct and control an organization regarding risk.

The general purpose is to reduce an industrial system's vulnerability as required in order to increase its resilience as justified considering current legal and/or regulatory requirements regarding the results of cost–benefit analyses. Relevant protection measures should be proposed that adequately safeguard and enable an organization to prevent or reduce the impact and consequences of potential disruptions.

After a major disruption, the recovery process is to be undertaken in order to restore system operation in a timely manner and improve activities, operations, facilities, and other key determinants of the affected organization where appropriate in order to increase its business resilience for the future.

2.6. BCM in Energy Companies

There have been various approaches proposed to apply the BCM concept in industrial practice. The standard BS 25999-1 [29] provided a proposal based on the concept of good practice. It was intended for use by anyone with responsibility for business operations or the provision of services, from top management through all levels of the organization. It was in principle foreseen for a single-site BCM or, with a more global presence, ranging from a sole trader through a small-to-medium enterprise (SME) to a large company employing thousands of people. However, this standard was withdrawn and replaced in industrial practice in favor of the ISO/DIS 22301 [4], which describes the basic requirements to be assigned in developing modern BCM systems.

As previously mentioned, BCM includes the recovery, management, and continuation of business activities in situations of business disruption as well as integrated management of the overall program through training, exercises, audits, and reviews to ensure that the business continuity plan stays current and up to date [48–50].

When analyzing energy and industrial companies, including their control systems [15,51,52], it is suggested that the following categories of potential disruptions be considered:

- Failures in logistics chains, delays in delivery of raw materials or semi-finished products by business partners, and/or delays in providing services, spare parts etc.
- Failures in electric energy distributed systems
- Power transformer station failures fires, cyberattacks, etc.
- Physical or cyberattack
- Failures and outages of ICT and CT (cloud technology) systems and networks designed using wired and/or wireless technology
- Failures and outages of OT systems and networks, including production lines and storage, and/or malfunctions of industrial automation and control systems (IACS)
- Extreme environmental phenomena, lightning storms, heavy rain, local flooding, flood, hurricane, or tornado, extremely high or low temperature, and heavy snowfall or icing
- Disturbances in critical infrastructure objects and systems needed to deliver water, electricity, gas etc.
- Fire or explosion
- Extreme emission of pollutants and/or dangerous substances
- Destruction due to potentially critical events in physical surroundings or infrastructure installations
- Earthquake and/or tsunami (at sites close to the shore)
- Sabotage, terrorism, or cyberterrorism against critical infrastructure objects/systems inspired by an external principal or agent
- Legislative changes

Only selected categories of potential disruptions will be discussed in the presented approach.

The consequences of an incident may vary significantly and can be far-reaching, including major accidents with both internal and external losses. These consequences might involve loss of life, environmental losses, and loss of assets or income due to the inability to deliver products and services on which the organization's strategy, reputation, or even economic survival might depend.

The importance of shaping the organizational culture and related safety and security culture is essential. It is a fundamental prerequisite both of successful activities and of avoiding failures in any organization, including a modern industrial company present within a competitive market.

Expected outcomes of an effective BCM program implemented in an energy or industrial company are as follows [4,49]:

- Key products and services are identified and protected, ensuring their continuity;
- Incident management capability is enabled to provide an effective response;
- The company understands its relationships with cooperating companies/organizations, relevant regulators and authorities, and emergency services;
- Staff are trained to respond effectively to an incident or disruption through appropriate exercises;
- Stakeholders' requirements are understood and able to be delivered;
- Staff receive adequate support and communications in the event of a disruption;
- The company's supply chain is better secured;
- The organization's reputation is protected and remains compliant with its legal and regulatory obligations.

In the energy sector, it is crucial to have maintain the operation of infrastructure equipment. This is supported by the correct application of BCM. As previously mentioned, there are many factors affecting the operation of any plant, including a power sector plant. These various factors are multidisciplinary and can be applied to different industry sectors.

Several indicators are used for decision-making in BCM, for instance [48], RTO (recovery time objective), the recovery time of a process or the required resources, and MTPD (maximum tolerable period of disruption), the maximal tolerable downtime which, when exceeded, seriously threatens the medium-term or long-term survival of the process or the organization. The maximum time for recovery (RTO) must be smaller than the maximum tolerable period of disruption (MTPD).

A formal set of procedures should be established to deal with information security incidents and identified weaknesses, which may have a physical component. This should encompass [44,49,50]:

- Detection of all information security incidents (and weaknesses) and related escalation procedures and channels;
- Reporting and logging of all information security incidents and weaknesses;
- Logging all responses and preventive and corrective actions taken;
- Periodic evaluation of all information security incidents and weaknesses;
- Learning from reviews of information security incidents(and weaknesses and making improvements to security and to the information security incident and weakness management scheme.

Service providers should ensure that all ICT systems essential for disaster recovery are tested regularly to ensure their continuing capability to support DR plans. Tests should be conducted whenever there are any significant changes in organizational requirements and/or changes in service provider capacity and capability that affect services to organizations. Examples of such changes include relocation of DR sites, major upgrades of ICT systems, and commissioning of new ICT systems.

There is an IT infrastructure in the energy sector, and problems with its proper operation contribute to power outages; information transmission deficiencies can cause blackouts in certain cases.

Several sets of various characteristics influencing performance and key performance indicators (KPIs) are listed [11] for use in evaluations and audits within the BCM of the industrial plant to support relevant decisions. Recovery Point Objective (RPO) and Recovery Time Objective (RTO) are two of the most important parameters of a disaster recovery or data protection plan. The RPO and RTO, along with a business impact analysis, provide the basis for identifying and analyzing viable strategies for inclusion in the business continuity plan of the BCM in relation to the previously discussed standards [4,48,50].

An objective of the recovery target time can be set, for instance, in the following cases:

- Resumption of product or service delivery after an incident, or resumption of a performance activity after an incident;
- Recovery of the ICT (information and communication technology) system or computer application after an incident, such as a hacker attack, or IT-OT system failure or

functional abnormality, such as abnormal performance of the industrial automation and control system (IACS).

The BCM approach outlined above is based on esteemed reports and international standards, including current legal and regulatory requirements.

The energy sector is critical to the operation of everything from households to critical infrastructure. In the current consideration of BCM, there are no specific explicit requirements for a particular sector, including the energy sector.

3. Proposed Integrated Functional Safety and Cybersecurity Evaluation in the Framework of BCM

In the approach presented below, current research issues are considered from the general perspective of BCM regarding the dependability, safety, and security of the ICT and ICS, including the SCADA (supervisory control and data acquisition) system. Their required functionality and architectures are discussed, distinguishing between information technology (IT) and operational technology (OT) in relevant systems and networks [40]. These systems require effective convergence for advanced manufacturing functionality and improved effectiveness in the realization of advanced manufacturing and business-related processes. In this article, an approach is proposed for integrated functional safety and cybersecurity analysis and management over the whole life cycle based on determining and verifying the safety integrity level (SIL) of the safety-related ICS system regarding the security assurance level (SAL) assigned to the relevant security domain.

The main objective is to outline a conceptual framework for including the above-mentioned technologies and systems within business continuity management activity. The proposed holistic management process identifies potential hazards and threats to an organization and their impact on manufacturing and business processes that, if realized, might cause disruptions and related losses. The purpose of the system is to provide a framework for building organizational resilience and preparing effective response, as such safeguards are important for company owners, key stakeholders, regulators, and local authorities [29,48–50] as well as crucial for brand, reputation, and value-creating activities.

The proposed BCM framework emphasizes the significance of a business continuity plan (BCP) for industrial companies (Figure 5).

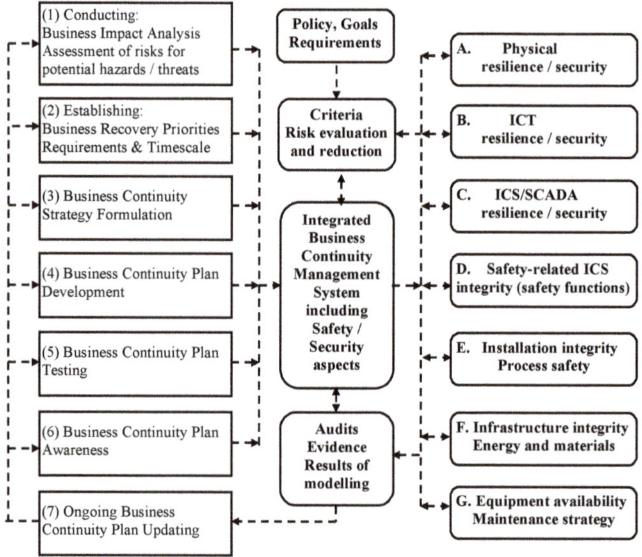

Figure 5. Proposed BCM framework for business continuity planning in industrial companies.

The left side of Figure 5 consists of seven specified discrete stages adapted from the standards in [44,50]; these are aimed at developing a comprehensive business continuity plan that will meet a company's business requirements, including the service providers. This is useful in developing recovery procedures (RP) for abnormal situations, failure events, or disaster recovery plans (DRP) [44] for cases of major disruptions and potential disasters.

In the middle part of this figure, basic elements of the approach to integrated BCM are specified, including the dependability, safety, and security aspects. The management activities are based on domain knowledge, current information, evidence, and results of modelling in the following areas:

- Formulating policies, goals, and domain, including legal and regulatory requirements and relevant standards and publications of good practice;
- Criteria for risk evaluation and reduction concerning dependability, safety, and security aspects, including domain key performance indicators (KPIs);
- Updated evidence, results of audits in design and plant operation, and results of modelling to support relevant decisions.

Audits can be (1) a first-party audit using internal resources, (2) a second-party audit initiated by a supplier, customer, contractor, and/or insurer, or (3) a third-party audit performed by an independent body against a recognized standard, i.e., ISO 9001.

On the right side of Figure 5, seven areas are specified and proposed by the authors for inclusion in the process of business continuity planning for a modern plant that requires relevant technical and organizational solutions in the following areas:

A. Physical resilience and security of company resources and assets;
B. Information and communication technology (ICT) resilience and security management over the whole life cycle;
C. Adequate resilience and security of the industrial automation and control system (IACS) and supervisory control and data acquisition (SCADA) system in a specific industrial network/domain and required security assurance level (SAL) [26];
D. Safety-related control systems designed and operated according to the functional safety concept with the required safety integrity level (SIL) [27];
E. Industrial installations and processes with the required physical and functional protection measures;
F. Infrastructure integrity for delivery of raw materials and energy (electricity, gas, oil) needed for production processes;
G. Equipment reliability/availability adequately maintained according to the strategy developed to achieve, for instance, a satisfactory level of overall equipment effectiveness (OEE).

These systems and networks require special attention during the design and operation of Industry 4.0 manufacturing systems due to their complexity, advanced functionality, and external communications. Their architectural complexity and openness make them susceptible to malfunctions and failures as well as vulnerable to external cyberattacks. According to published data, the probability of such attacks on various industrial systems and networks in most European countries is relatively high.

Due to the scope of the problems outlined above, only selected issues will be discussed. In the following sections fundamental aspects related to the Industry 4.0 concept are presented, namely, ICT systems and networks (B in Figure 5), ICS/SCADA resilience and security (C), and safety-related ICS (D) designed for implementing the defined safety functions [27,42,53,54] of the required safety integrity level (SIL) of a safety function in order to reduce relevant risks. The determined SIL is then verified using a probabilistic model of the safety-related ICS of the architecture, including communication conduits.

To better illustrate the authors' new approach, Figure 6 shows five framework elements that directly extend the BCM process.

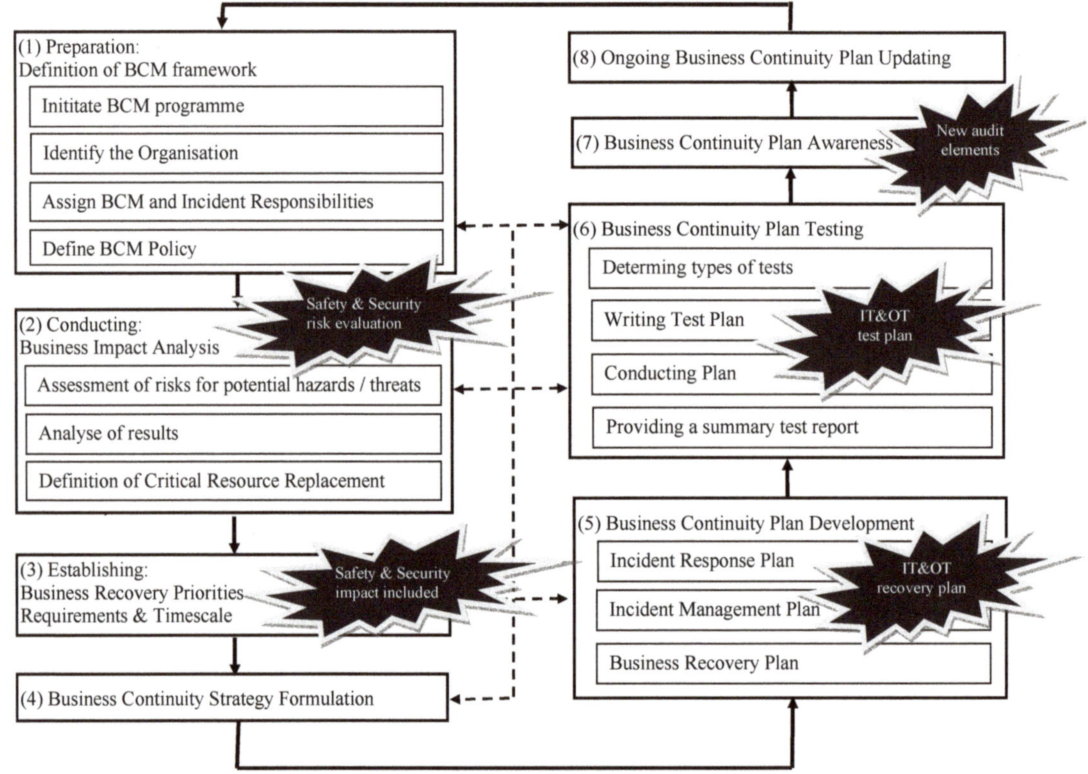

Figure 6. Impact of the proposed framework on the BCM process.

The first element of the new approach is to first incorporate the safety and security aspects discussed above into the risk analysis and then throughout the Business Impact Analysis process. The aim of information security management (ISM) is to fulfill specified requirements concerning the CIA triad (Figure 4) of the ICT systems regarding information storage, transfer, and related services. When an organization implements an ISMS (information security management system), the risks of interruptions to business activities for any reason should be identified and evaluated [20,55].

In the third step of the BCM process (Establishing), the conclusions of Step 2 should be considered, including new safety and cybersecurity aspects.

During development of the Business Continuity Plan, the dependencies of IT on OT and their impact on functional safety must first be considered; second, the impact of these events on the recovery plan must be assessed. Planning for business continuity, fallback arrangements for information processing, and communication facilities become beneficial during periods of minor outages and are essential for ensuring information and service availability during a major failure or disaster that requires complete and effective recovery of activities over a period of time.

The fourth important link in the proposed framework is the inclusion of aspects of the risk analysis and the prepared recovery plan in the process of periodic testing and verification.

The last new element appears in the final two steps of the BCM cycle. As previously mentioned, audits are of key importance in any management system, especially in a hazardous industrial plant. Previous authors have examined audit documentation prepared and used by an industrial company as part of a third-party audit in a refinery concerning the design and operation of safety-related ICS in relation to defined generic and plant criteria [56]. The audit results and conclusions were then discussed with the staff responsible for functional safety to further mitigate risks by implementing the indicated technical and organizational solutions. An important objective in implementing a BCM in a hazardous plant is to satisfy the expectations of stakeholders and insurance companies [10,11] in order to assure a satisfactory level of business continuity, safety, and security. This can be achieved thanks to the implementation in industrial practice of advanced, consistent and effective BCM systems.

Thus, the BCM is useful in taking a systemic and proactive approach to dealing with dependability, safety, and security issues. It specifies various interrelated process-based activities and procedures for the identification of hazards and threats in order to evaluate relevant risks, supporting safety and security-related decision-making in changing conditions and over the whole plant life cycle.

4. Case Study

4.1. Safety Aspects

The risk analysis phase of a plant's BCM takes into account the continuity of the media supply, which is directly linked to the plant's gas boiler room. As part of the functional safety and cybersecurity risk analysis, analyses were performed as a basis for this risk analysis. In this example, only one of the safety functions is presented. A safety function of high-pressure monitoring operating in the low demand mode in a process installation is presented. The high pressure of the steam in the process loop provokes the safety function to drop power to a pair of solenoid valves, which leads to venting to a pneumatic actuator, placing a pair of valves into their failsafe position. From the risk evaluation, the safety integrity level of this function was determined to be SIL 3. The safety function to be implemented in the safety-related ICS architecture is shown in Figures 7 and 8. The related at BCM framework, including the safety and security aspects, is shown in Figure 9.

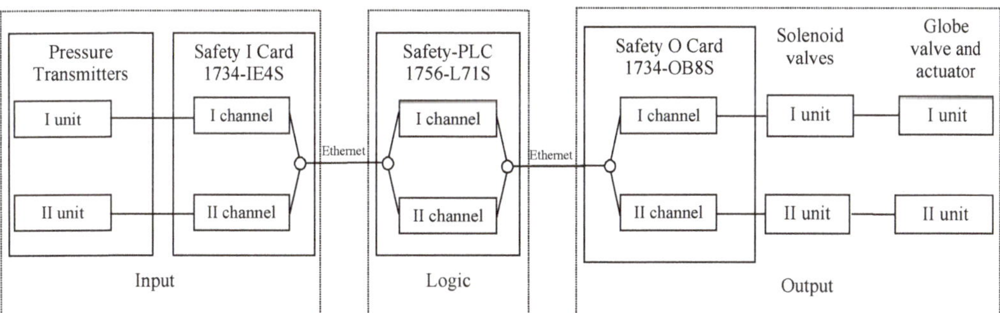

Figure 7. The architecture of the ICS system with implemented safety function.

In the analysed example, the 4–20 mA two-wire pressure transmitters are directly wired into analog input modules. The safety controller and the input and output cards are connected on an EtherNet/IP network. The final control elements of this safety function are the combination of solenoids, actuators, and globe valves. The controller and safety I/O modules have a built-in HFT = 1 (two field signals are used). The sensors and final elements require redundant hardware in the 1oo2 configuration to meet the required HFT = 1. The data for evaluating the probability of failure on demand average PFD$_{avg}$ of subsystems was calculated by the authors based on data provided by manufacturers of the components (Table 4).

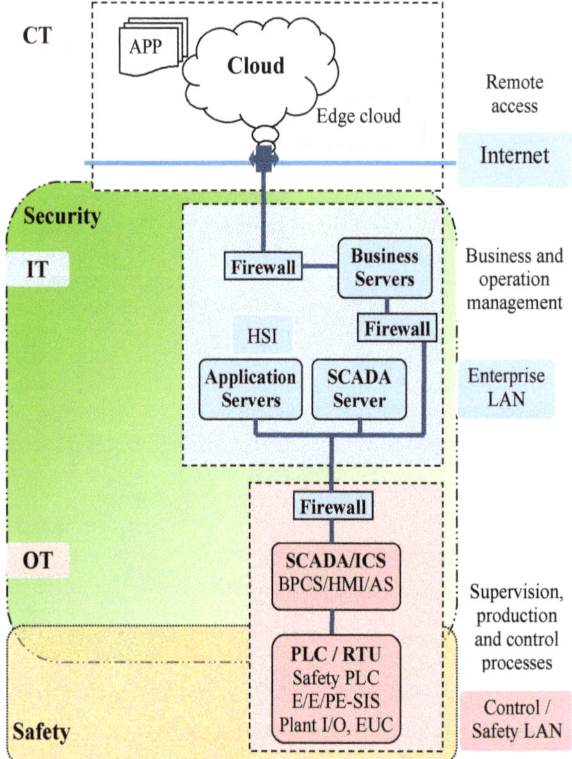

Figure 8. Analysed object architecture.

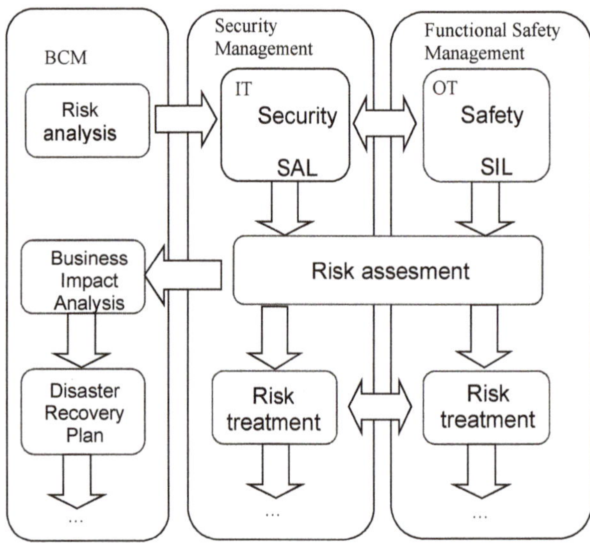

Figure 9. Diagram of relations of BCM framework, including safety and security aspects.

Table 4. Reliability data for safety-related ICS components for implementing the safety function.

Subsystem	SIL	PFD_{avg}
A. Input subsystem Pressure transmitter Analog Input Card	SIL 4	3.1×10^{-5}
B. Logic subsystem Safety PLC	SIL 4	3.5×10^{-5}
C. Output subsystem Digital Output Card Solenoid valve Globe valve & Pneumatic Actuator	SIL 4	4.6×10^{-5}

The value of PFD_{avg} for the considered safety-related ICS is calculated from the formula [39]:

$$PFD_{avg} \cong PFD_{avg}^{A} + PFD_{avg}^{B} + PFD_{avg}^{C} \qquad (4)$$

Thus, in this case study, $PFD_{avg} \cong 11.2 \times 10^{-5}$; the safety integrity level of SIL 3 was obtained via the results of probabilistic modelling, with the interval criteria presented in the second column of Table 1 and the architecture constraints presented in the IEC 61508 series standard.

4.2. Safety-Related ICS Aspects

Considering the domain of the safety-related ICS in which the safety function was implemented, including the communication conduits, the SL-A vector was evaluated as follows: (3 4 3 3 3 3 4). Assuming that weights of all SL_i are equal ($w_i = 1/7$), using Equation (3) the obtained result is $SI^{Do} = 3.28$ and the assigned security assurance level is SAL 3. From column 4 of Table 3, the final safety integrity level validated regarding the security aspects in the domain of interest is SIL 3, the same as required. Therefore, in this case there is no need to propose improvements to the safety-related system [40]. If the SAL obtained for another less secure domain was lower, e.g., SAL 2, then the assigned safety integrity level should be lower, i.e., SIL 2 (see Table 3).

4.3. Risk Treatment

From a risk management point of view, it would be justified to consider changing the configuration of the sensor subsystem shown in Figure 7 from 1oo2 to 2oo3 in order to reduce the probability of spurious operation of this safety-related ICS. It is known that while the 2oo3 configuration has a slightly higher PFD_{avg}, it has a much lower probability of spurious operation than configuration 1oo2. Probabilistic modeling of the safety-related ICS consisting of the 2oo3 configuration, including the influence of CCFs and the architectural constrains on subsystems regarding their HFT and S_{FF}, is described in detail in [25,43].

4.4. Business Continuity Management Impact

Based on the information previously mentioned in the example above, the team can assess the impact of system architecture and functional and cybersafety safeguards on the criticality of the gas boiler, which translates into the BCM of the entire plant. This in turn allows the enterprise to engage in Business Continuity Planning-wide planning, e.g., creating the capacity to produce a range of products in several factories.

The next stage of BCM is to create a business recovery plan that includes both IT and OT infrastructure. As IT practices are well known, we omit the related description. As far as OT is concerned, especially in terms of functional safety, it is necessary to highlight the creation of backup programs of drivers and safety drivers, knowledge of firmware versions of devices of control systems, protection of spare parts of control elements which can be destroyed or infected, description of procedures verifying damage, and procedures allowing for the restarting of production after replacement of damaged or infected elements.

The final stage of the BCM process, enriched with new analysis elements, is the test plan. At this stage, it is necessary to equip maintenance personnel with appropriate procedures and instructions to test the disaster event and production recovery in a safe way for the continuity of production in the scope resulting from the risk analysis enriched with functional safety elements for OT and IT.

The frequency of performing a backup depends directly on the Recovery Point Objective (RPO) indicator assumed during the analysis. However, the size of the stock of key spare parts depends on the adopted Recovery Time Objective (RTO)

4.5. Summary

This example demonstrates that in a modern industrial plant equipped with both safety functions and IT networks, these two functionalities intermingle and create interactions that have a direct impact on BCM analyses. Their consideration is essential for a comprehensive analysis of all risks and the creation of an appropriate action plan.

5. Conclusions

In this article, an integrated functional safety and cybersecurity evaluation approach is proposed in a framework for business continuity management (BCM) to deal systematically with vulnerabilities that could influence an industrial plant's dependability, safety, and security. Industrial energy companies, including those using Industry 4.0 business and technical solutions, have to pay attention to shaping their resilience regarding existing and emerging hazards and threats, including cyberattacks. This issue concerns the energy sector, power plants, and distributed renewable energy stations.

In such energy plants, information and communication technologies (ICT) and industrial automation and control systems (IACS) play important roles. Using advanced technologies in modern energy manufacturing systems and processing plants can result in management impediments due to their openness to external systems and networks through various communication channels. This makes company assets and resources potentially vulnerable to risk, e.g., due to cyberattacks. In the BCM-oriented approach proposed here, both preventive and recovery activities are considered in light of engineering best practices and following suggested selected international standards, reports, and domain publications.

Potential impediments in energy industrial practice have been identified related to OT security when this technology consists of devices (hardware and software) from several different producers/suppliers. This can cause substantial difficulties in pathing software within relevant computer systems and networks. Thus, this issue requires special attention during the design, implementation, and maintenance of business continuity management systems.

The dependability and security of safety-related ICS in which defined safety functions are implemented can be influenced by both technical and organizational factors. These are related to the quality and reliability of hardware and software. These aspects require further research, especially in the context of the design and operation of highly complex hazardous industrial installations and their ICS, as these must be designed with regard to the defense in depth concept when justified in the context of the risk evaluation results obtained.

Traditionally, manufacturing installations include both information technology (IT) and operational technology (OT). More recently, cloud technology (CT) is often considered to improve data transfer and storage in the context of business management in distributed Industry 4.0 companies.

Advanced automation and control systems are currently in development, based, for instance, on the open platform communication unified architecture (OPC UA) protocol for improved network scalability and implementing new AutomationML concepts [49]. These technologies enable advanced production flexibility and effectiveness. The IT, OT, and

IACS, including safety-related ICS, can be considered more generally as a cyber-physical system (CPS). Additional research should be undertaken in order to deal systematically with distributed co-operating manufacturing systems, including their dependability, safety, and security aspects, regarding an advanced BCM system for improving effectiveness and resilience over the whole plant life cycle. Our future research work will further develop BCM topics in the energy sector related to hydrogen storage and renewable energy technologies. With respect to these topics it is extremely important to include an integrated approach to functional safety and cybersecurity analysis.

Author Contributions: Conceptualization, K.T.K.; methodology, K.T.K. and M.Ś.; validation, J.P. and E.P.; formal analysis, E.P., J.P. and M.Ś.; investigation, K.T.K., M.Ś., J.P. and E.P.; resources, E.P. and M.Ś.; writing—original draft preparation, K.T.K.; writing—review and editing, E.P., J.P. and M.Ś.; visualization, K.T.K.; supervision, M.Ś., J.P. and E.P. All authors have read and agreed to the published version of the manuscript.

Funding: This research was supported by Gdańsk University of Technology.

Institutional Review Board Statement: Not applicable.

Informed Consent Statement: Not applicable.

Data Availability Statement: Not applicable.

Conflicts of Interest: The authors declare no conflict of interest.

References

1. SIEMENS Industrial Security. Available online: https://new.siemens.com/global/en/products/automation/topic-areas/industrial-security.html (accessed on 10 June 2021).
2. Abdo, H.; Kaouk, M.; Flaus, J.M.; Masse, F. Safety and Security Risk Analysis Approach to Industrial Control Systems. *Comput. Secur.* **2018**, *72*, 175–195. [CrossRef]
3. Li, S.W. Architecture Alignment and Interoperability, an Industrial Internet Consortium and Platform Industry 4.0. Available online: https://www.iiconsortium.org/pdf/JTG2_Whitepaper_final_20171205.pdf (accessed on 10 June 2021).
4. *ISO/DIS 22301*; Security and Resilience—Business Continuity Management Systems—Requirements. International Organization for Standardization: Geneva, Switzerland, 2019.
5. Xing, J.; Zio, E. *An Integrated Framework for Business Continuity Management of Critical Infrastructures*; CRC Press: Boca Raton, FL, USA, 2016; pp. 563–570.
6. Lundteigen, M.A.; Rausand, M.; Utne, I.B. Integrating RAMS engineering and management with the safety life cycle of IEC 61508. *Reliab. Eng. Syst. Saf.* **2009**, *94*, 1894–1903. [CrossRef]
7. Saraswat, S.; Yadava, G.S. An overview on reliability, availability, maintainability and supportability (RAMS) engineering. *Int. J. Qual. Reliab. Manag.* **2008**, *25*, 330–344. [CrossRef]
8. Misra, K.B. (Ed.) *Handbook of Advanced Performability Engineering*; Springer Nature: Cham, Switzerland, 2021.
9. Niemimaa, M. Interdisciplinary Review of Business Continuity from an Information Systems Perspective: Toward an Integrative Framework. *Commun. Assoc. Inf. Syst.* **2015**, *37*, 4. [CrossRef]
10. Gołębiewski, D.; Kosmowski, K. Towards Process-Based Management System for Oil Port Infrastructure in Context of Insurance. *J. Pol. Saf. Reliab. Assoc.* **2017**, *8*, 23–37.
11. Kosmowski, K.T.; Gołębiewski, D. Functional Safety and Cyber Security Analysis for Life Cycle Management of Industrial Control Systems in Hazardous Plants and Oil Port Critical Infrastructure Including Insurance. *J. Pol. Saf. Reliab. Assoc.* **2019**, *10*, 99–126.
12. Kosmowski, K.T. Systems engineering approach to functional safety and cyber security of industrial critical installations. In *Safety and Reliability of Systems and Processes*; Kołowrocki, K., Bogalecka, M., Dąbrowska, E., Torbicki, M., Eds.; Gdynia Maritime University: Gdynia, Poland, 2020; pp. 135–151.
13. *Systems Engineering Fundamentals*; Defense Acquisition University Press: Fort Belvoir, VA, USA, 2001.
14. Białas, A. *Semiformal Common Criteria Compliant IT Security Development Framework*; Studia Informatica; Silesian University of Technology Press: Gliwice, Poland, 2008.
15. Kriaa, S.; Pietre-Cambacedes, L.; Bouissou, M.; Halgand, Y. Approaches Combining Safety and Security for Industrial Control Systems. *Reliab. Eng. Syst. Saf.* **2015**, *139*, 156–178. [CrossRef]
16. CISA Assessments: Cyber Resilience Review. Available online: https://www.cisa.gov/uscert/resources/assessments (accessed on 10 February 2020).
17. Leitão, P.; Colombo, A.W.; Karnouskos, S. Industrial Automation Based on Cyber-Physical Systems Technologies: Prototype Implementations and Challenges. *Comput. Ind.* **2016**, *81*, 11–25. [CrossRef]

18. MERGE. Safety & Security, Recommendations for Security and Safety Co-Engineering, Multi-Concerns Interactions System Engineering. ITEA2 Project No. 11011. Available online: https://itea4.org/project/workpackage/document/download/2837/D3.4.4.%20MERgE%20-%20Recommendations%20for%20Security%20and%20Safety%20Co-engineering%20v3%20partA.pdf (accessed on 1 June 2021).

19. Integrated Design and Evaluation Methodology. Security and Safety Modelling; Artemis JU Grant Agr., No. 2295354. Available online: http://sesamo-project.eu/sites/default/files/downloads/publications/integrated-design-and-evaluation-communication-material.pdf (accessed on 5 June 2018).

20. Boehmer, W.J. Survivability and business continuity management system according to BS 25999. In Proceedings of the IEEE 3rd International Conference on Emerging Security Information, Systems and Technologies, Athens, Greece, 18–23 June 2009; Volume 1, pp. 142–147.

21. Zawiła-Niedźwiecki, J. *Operational Risk Management in Assuring Organization Operational Continuity*; Edu-Libri.: Kraków, Poland, 2013. (In Polish)

22. Cyber Security for Industrial Automation and Control Systems, Health and Safety Executive (HSE) Interpretation of Current Standards on Industrial Communication Network and System Security, and Functional Safety 2015. Available online: https://www.hse.gov.uk/foi/internalops/og/og-0086.pdf (accessed on 5 May 2021).

23. Kosmowski, K.T. Functional safety and cybersecurity analysis and management in smart manufacturing systems. In *Handbook of Advanced Performability Engineering*; Krishna, B.M., Ed.; Springer Nature: Cham, Switzerland, 2021.

24. Kościelny, J.M.; Syfert, M.; Fajdek, B. Modern Measures of Risk Reduction in Industrial Processes. *J. Autom. Mob. Robot. Intell. Syst.* **2019**, *1*, 20–29. [CrossRef]

25. Kosmowski, K.T. *Functional Safety and Reliability Analysis Methodology for Hazardous Industrial Plants*; Gdansk University of Technology Publishers: Gdańsk, Poland, 2013.

26. *IEC 62443*; Security for Industrial Automation and Control Systems. Parts 1–14 (Some Parts in Preparation). The International Electrotechnical Commission: Geneva, Switzerland, 2018.

27. *IEC 61508*; Functional Safety of Electrical/ Electronic/ Programmable Electronic Safety-Related Systems, Parts 1–7. The International Electrotechnical Commission: Geneva, Switzerland, 2016.

28. Gabriel, A.; Ozansoy, C.; Shi, J. Developments in SIL Determination and Calculation. *Reliab. Eng. Syst. Saf.* **2018**, *177*, 148–161. [CrossRef]

29. *BS 25999-1*; Business Continuity Management—Part 1: Code of Practice. British Standard Institution: London, UK, 2006.

30. *SP 800-82r2*; Guide to Industrial Control Systems (ICS) Security. NIST: Gaithersburg, MD, USA, 2015.

31. *ETSI TS 102 165-1*; CYBER Methods and Protocols. Part 1: Method and pro Forma for Threat, Vulnerability, Risk Analysis (TVRA). Technical Specs; ETSI: Sophia Anthipolis, France, 2017.

32. Kosmowski, K.T.; Śliwiński, M. Organizational culture as prerequisite of proactive safety and security management in critical infrastructure systems including hazardous plants and ports. *J. Pol. Saf. Reliab. Assoc.* **2016**, *7*, 133–146.

33. ISA. *Security of Industrial Automation and Control Systems, Quick Start Guide: An Overview of ISA/IEC 62443 Standards*; ISA—International Society of Automation: Alexander, NC, USA, 2020.

34. Saleh, J.H.; Cummings, A.M. Safety in the Mining Industry and the Unfinished Legacy of Mining Accidents. *Saf. Sci.* **2011**, *49*, 764–777. [CrossRef]

35. Subramanian, N.; Zalewski, J. Quantitative Assessment of Safety and Security of System Architectures for Cyberphysical Systems Using NFR Approach. *IEEE Syst. J.* **2016**, *2*, 397–409. [CrossRef]

36. *IEC 61511*; Safety Instrumented Systems for the Process Industry Sector. Parts 1–3. The International Electrotechnical Commission: Geneva, Switzerland, 2016.

37. Holstein, D.K.; Singer, B. *Quantitative Security Measures for Cyber & Safety Security Assurance*; ISA: Alexander, NC, USA, 2010.

38. Śliwiński, M.; Piesik, E.; Piesik, J. Integrated Functional Safety and Cybersecurity Analysis. *IFAC Pap. OnLine* **2018**, *51*, 1263–1270. [CrossRef]

39. *IEC 62061*; Safety of Machinery—Functional Safety of Safety-Related Electrical, Electronic, and Programmable Electronic Control Systems. The International Electrotechnical Commission: Geneva, Switzerland, 2018.

40. Kosmowski, K.T.; Śliwiński, M.; Piesik, J. Integrated Functional Safety and Cybersecurity Analysis Method for Smart Manufacturing Systems. *TASK Q.* **2019**, *23*, 1–31.

41. *IEC 63074*; Security Aspects Related to Functional Safety of Safety-Related Control Systems. The International Electrotechnical Commission: Geneva, Switzerland, 2017.

42. Braband, J. What's security level got to do with safety integrity level? In Proceedings of the 8th European Congress on Embedded Real Time Software and Systems, Toulouse, France, 27–29 January 2016.

43. Kosmowski, K.T. Safety integrity verification issues of the control systems for industrial power plants. In Proceedings of the International Conference on Diagnostics of Processes and Systems, Sandomierz, Poland, 11–13 September 2017; pp. 420–433.

44. *ISO/IEC 24762*; Information Technology—Security Techniques—Guidelines for Information and Communications Technology Disaster Recovery Services. International Organization for Standardization: Geneva, Switzerland, 2008.

45. *ISO/DTR 22100*; Safety of Machinery—Guidance to Machinery Manufacturers for Consideration of Related IT Security (Cyber Security) Aspects. International Organization for Standardization: Geneva, Switzerland, 2018.

46. *IEC TR 63074*; Safety of Machinery—Security Aspects to Functional Safety of Safety-Related Control Systems. The International Electrotechnical Commission: Geneva, Switzerland, 2019.
47. *ISO/IEC 27005*; Information Technology—Security Techniques—Information Security Risk Management. International Organization for Standardization: Geneva, Switzerland, 2018.
48. *BSI-Standard 100-4*; Business Continuity Management. Federal Office for Information Security (BSI): Berlin, Germany, 2009.
49. *ISO/PAS 22399*; Societal Security—Guideline for Incident Preparedness and Operational Continuity Management. International Organization for Standardization: Geneva, Switzerland, 2007.
50. *ISO/IEC 27031*; Information Technology—Security Techniques—Guidelines for Information and Communication Technology Readiness for Business Continuity. International Organization for Standardization: Geneva, Switzerland, 2011.
51. Kanamaru, H. Bridging functional safety and cyber security of SIS/SCS. In Proceedings of the IEEE 56th Annual Conference of the Society of Instrument and Control Engineers of Japan, Kanazawa, Japan, 19–22 September 2017.
52. Smith, D.J. *Reliability, Maintainability and Risk. Practical Methods for Engineers*, 9th ed.; Butterworth-Heinemann: Oxford, UK, 2017.
53. Piesik, E.; Śliwiński, M.; Barnert, T. Determining the Safety Integrity Level of Systems with Security Aspects. *Reliab. Eng. Syst. Saf.* **2016**, *152*, 259–272. [CrossRef]
54. Kosmowski, K.T.; Śliwiński, M. *Knowledge-Based Functional Safety and Security Management in Hazardous Industrial Plants with Emphasis on Human Factors*; Advanced Control and Diagnostic Systems; PWNT: Gdańsk, Poland, 2015.
55. Felser, M.; Rentschler, M.; Kleinberg, O. Coexistence standardisation of operational technology and information technology. *Proc. IEEE* **2019**, *104*, 962–976. [CrossRef]
56. Rogala, I.; Kosmowski, K.T. *Audit Document Concerning Organizational and Technical Aspects of the Safety-Related Control System Design and Operation at a Refinery (Access Restricted)*; Automatic Systems Engineering, Gdańsk and Gdańsk University of Technology: Gdańsk, Poland, 2012.

Article

Multi-Objective Optimization via GA Based on Micro Laser Line Scanning Data for Micro-Scale Surface Modeling

J. Apolinar Muñoz Rodríguez

Centro de Investigaciones en Óptica, A. C., Lomas del Bosque 115, Col. Comas del Campestre, Leon 37000, GTO, Mexico; munoza@cio.mx

Abstract: Industry 4.0 represents high-level methodologies to make intelligent, autonomous, and self-adaptable manufacturing systems. Additionally, the surface modeling technology has become a great tool in industry 4.0 for representing the surface point cloud. Thus, the micro-scale machining technology requires efficient models to represent micro-scale flat and free-form surfaces. Therefore, it is fundamental to perform surface modeling through artificial intelligence for representing small surfaces. This study addressed multi-objective optimization via genetic algorithms and micro laser line projection to accomplish surface models for representing micro-scale flat and free-form surfaces, where an optical microscope system retrieves micro-scale topography via micro laser line coordinates and the multi-objective optimization constructs the flat and free-form surface models through genetic algorithms and micro-scale topography. The multi-objective optimization determines the surface model parameters through exploration and exploitation, and the solution space is deduced via surface data. The surface model generated through the multi-objective optimization fit accurately to the micro-scale target surface. Thus, the proposed technique enhanced the fitting of micro-scale flat and free-form surface models, which were deduced via gray-level images of an optical microscope. This enhancement was validated by a discussion between the multi-objective optimization via genetic algorithms and the micro-scale surface modeling via optical microscope imaging systems.

Keywords: micro-scale surface modeling; micro laser line contouring; multi-objective optimization; optical microscope imaging

Citation: Rodríguez, J.A.M. Multi-Objective Optimization via GA Based on Micro Laser Line Scanning Data for Micro-Scale Surface Modeling. *Energies* **2022**, *15*, 6571. https://doi.org/10.3390/en15186571

Academic Editor: Marko Mladineo

Received: 21 July 2022
Accepted: 24 August 2022
Published: 8 September 2022

1. Introduction

Nowadays, the manufacturing industry is developing intelligent tools to accomplish the requirements of industry 4.0. The intelligent tools include cyber-physical systems, Internet of Things, information technology, digital manufacturing activities, and so on [1]. The cyber-physical systems are involved in the manufacturing systems based on artificial intelligence and the self-adaptable process [2]. However, the Internet of Things provides tools to store and share the surface point cloud in the manufacturing systems during the production process [3]. In this way, the cyber-physical systems, Internet of Things, digital manufacturing, and information technology determine the surface point cloud and perform surface modeling to represent a small object surface in digital form. Additionally, the surface modeling provides fundamental tools in industry 4.0 to represent computationally flat and free-form surfaces [4,5]. In the same way, the flat and free-form surface modeling at the micro-scale provides computational tools in industry 4.0 to represent micro-scale objects [6,7], where the computational tools are constructed via mathematical models to generate flat and free-form surfaces at the micro-scale. Actually, micro-scale flat and free-form surface modeling is performed via optical microscope systems based on gray-level image processing in industry 4.0 [8,9]. The microscope systems perform surface recovering and mathematical modeling to accomplish micro-scale surface modeling. Typically, the micro-scale surface modeling is accomplished through the least-squares method, and the surface data are determined via optical microscope images [10,11]. However, the surface fitting of the least-squares method produced inaccuracy with respect to the target surface [12].

Therefore, artificial intelligence methods have been implemented to perform micro-scale surface modeling in industry 4.0. For instance, the particle swarm has been implemented to perform free-form surface modeling through equations based on particle movement [13,14], where the parameter optimization has been carried out through a population generated by means of particle velocity. Additionally, particle swarm optimization has been employed in several optimization applications to achieve a good fit and convergence [15,16]. However, the particle swarm is a metaheuristic algorithm more employed to optimize surface modeling. Additionally, ant colony has been implemented to perform free-form surface modeling in industry 4.0 [17,18], where the parameter optimization has been carried out by selecting pathways generated by the ant pheromone to accomplish the computational model. In the same way, simulated annealing has been implemented to construct free-form surface models in industry 4.0 [19,20], where the model parameters have been computed through a perturbation in an equation system. Additionally, fuzzy logic has been implemented to generate free-form surface models in industry 4.0 [21,22], where the model parameters have been optimized by using contactless scanning of the transtibial prosthetic socket. Furthermore, virtual reality was implemented to perform free-form surface models through a spline function [23], where a convex approximation was carried out through the spline basis function to determine concave surface models. Moreover, machine learning was implemented to perform flat surface models via neural networks [24], where the flat surface equation was constructed from the topography contour data. On the other hand, the micro-scale surface models are constructed through surface data, which are determined by means of optical microscope images. For example, a microscope imaging system determined free-form surfaces at the micro-scale by computing data through the optical microscope images [25]. Additionally, a microscope imaging system performed flat contouring at the micro-scale by computing surface data through the optical microscope images [26], where a frequency transform was computed from the gray-level image to determine the surface data. Moreover, a microscope imaging system retrieved flat topography at the micro-scale through a regression model and optical microscope images [27], where a fitting function was generated by means of the gray-level image to determine the surface data. The above microscope arrangements perform the surface modeling at the micro-scale through the surface data, which are determined by means of optical microscope images. However, the gray-level profile does not depict the topography contour with the best accuracy. The profile inaccuracy is caused by the object reflectance, light transmitter source, and viewer direction. Therefore, the computational model produces flat and free-form surfaces away from the target surface. In addition, the above-mentioned methods optimize the surface model parameters through a search space, which is not deduced from the surface data. For instance, the free-form surface models generated through the simulated annealing, particle swarm, and ant colony are not accomplished by means of surface data. Such algorithms determine the model parameters through a solution space, which is not deduced from the surface topography. As a consequence of missing references, complicated procedures should be implemented to optimize the model parameters. Moreover, the minimization function is deduced through a mathematical expression, which includes additional variables to the surface model. Therefore, the additional variables should be computed to optimize the surface model parameters. Although the flat and free-form surface modeling at the micro-scale represents one of the central paradigms of industry 4.0, it still requires further research and development. Therefore, it is established that the flat and free-form surface modeling performed at the micro-scale via optical microscope systems still represents a difficult challenge. To enhance the flat and free-form surface modeling at the micro-scale, a multi-objective optimization based on several objective functions and surface contour data is required. It is because the free-form surface is generated through several surface functions. The multi-objective optimization via genetic algorithms was implemented in several applications [28] and provided good fitting and convergence results.

The proposed flat and free-form surface modeling at the micro-scale is performed through multi-objective optimization and micro laser line scanning, which retrieves surface contour with great accuracy. The multi-objective optimization is implemented through a genetic algorithm, which employs several objective functions to construct micro-scale flat and free-form surface models. In this way, the free-form surface modeling is implemented through objective functions, which are generated through the Bezier basis functions and surface control points. Additionally, the flat surface modeling is implemented through objective functions, which are generated through the plane equation and surface data. Thus, the genetic algorithm performs micro-scale flat and free-form surface modeling by employing surface coordinates, which are computed from the laser line coordinates. To carry it out, the multi-objective optimization determines the solution space from the surface coordinates. Then, exploration and exploitation are performed to determine the optimal model parameters from the solution space. The main approach to perform flat and free-form surface modeling through genetic algorithms is determined by the quality gap, number of iterations, and suitable structure. For instance, the particle swarm, ant colony, and simulated annealing do not determine the solution space through the surface data. Instead, the genetic algorithm determines the solution space from the surface data. This leads to obtaining an initial population near of the optimal solution and reduces the number of iterations. Furthermore, the other intelligent algorithms employ additional parameters to the surface model and increase the number of iterations. Additionally, the genetic algorithm provides a better surface model fitting than the other intelligent algorithms. Moreover, the genetic algorithm performs exploration and exploitation to find the optimal solution inside or outside of the best candidates. Based on these statements, the genetic algorithm was chosen to perform the flat and free-form surface models. In this way, the micro-scale flat and free-form surface modeling was performed by an optical microscope vision system on which a CCD camera and a 38 μm laser line were attached. Thus, the micro laser line was projected on the surface and the reflection depicted the surface contour, which was captured by the CCD camera. The micro-scale surface coordinates were computed through the microscope geometry and the laser line position. Thus, the micro-scale flat and free-form surface models were accomplished through the surface coordinates, which are not used by the optical microscope systems. Therefore, the micro-scale flat and free-form surface modeling via multi-objective optimization and the micro laser line scanning enhanced the surface model fitting of the optical microscope systems, where the fitting accuracy was deduced through the difference between the data computed through the surface model and the real surface data. The multi-objective optimization was employed to construct micro-scale models of rectangular surfaces. The viability of the multi-objective optimization via genetic algorithms was deduced through the model fitting, run time, algorithm structure, and results' accuracy. To corroborate this statement, a discussion is provided about the fitting accuracy of the surface models at the micro-scale. In this way, the viability of the proposed technique is corroborated by the fitting accuracy of the micro-scale flat and free-form surface modeling. The paper is organized as follows: the multi-objective optimization to construct the free-form surface model at the micro-scale is explained in Section 2.1, the multi-objective optimization to construct the flat surface model at the micro-scale is explained in Section 2.2, the surface contouring at the micro-scale through the microscope vision system is described in Section 2.3, the results of the flat and free-form surface modeling at the micro-scale are shown in Section 3, and the model fitting discussion is included in Section 4.

2. Materials and Methods

2.1. Multi-Objective Optimization for Micro-Scale Free-Form Surface Modeling

The free-form surface modeling at the micro-scale is performed through the multi-objective optimization and surface coordinates, which are computed via laser line scanning. Typically, the multi-objective optimization employs several objective functions to be minimized and guarantees to find the ideal solution [29]. In this way, the multi-objective optimization operates on a set of solution spaces and employs a moderated time to achieve the optimization. Additionally, exploration and exploitation are implemented by the multi-objective optimization to determine the optimal solution from the solution space. Thus, the free-form surface modeling at the micro-scale is carried out through the multi-objective optimization based on a genetic algorithm and the contour data. To do so, the free-form surface is retrieved via micro laser line scanning. The contour coordinates are shown in Figure 1, and they are represented by $(x_{i,j}, y_{i,j}, z_{i,j})$, where the sub-indices (i, j) are established in the x-axis and y-axis, respectively. In this way, the surface provides the coordinates $(z_{0,0}, z_{1,0}, \ldots, z_{n,0}, z_{n,1}, \ldots, z_{n,m})$ in the z-axis, where the sub-indices (n, m) depict the coordinate number in the x-axis and y-axis, respectively.

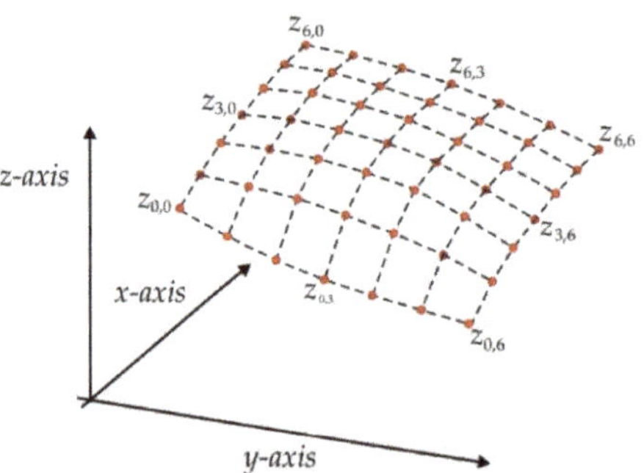

Figure 1. Free-form surface data to perform surface modeling.

From the surface coordinates, the free-form surface model is constructed through a bi-cubic Bezier surface [30], which is described by the expression

$$
S_{p,q}(u,v) = \sum_{s=0}^{s=3} \sum_{r=0}^{r=3} B_r(u) B_s(v) P_{i,j}, \quad u,v \in [0,1],
$$
$$
B_r(u) = \frac{3!}{r!(3-r)!}(1-u)^{3-r} u^r, \quad B_s(v) = \frac{3!}{s!(3-s)!}(1-v)^{3-s} v^s.
$$
(1)

In this equation, u is in the x-direction, v is in the y-direction, and $P_{i,j}$ are control points that move the Bezier surface toward the surface $z_{i,j}$. The sub-indices (i, j) are related to the sub-indices (r, s) through the terms $i = r + p*3$ and $j = s + q*3$, respectively. In this way, the free-form surface model is represented by the surfaces $S_{0,0}(u,v)$, $S_{1,0}(u,v)$, $\ldots$, $S_{n/3,0}(u,v)$, $\ldots$, $S_{n/3,m/3}(u,v)$. Thus, the free-form surface model is defined by the equation

$$
\begin{bmatrix} S_{0,0}(u,v) \\\\ S_{1,0}(u,v) \\\\ S_{0,1}(u,v) \\\\ \vdots \\\\ S_{n/3,m/3}(u,v) \end{bmatrix} =
\begin{bmatrix}
\begin{aligned} &B_{0,0}P_{0,0} + B_{1,0}P_{1,0} + B_{2,0}P_{2,0} + B_{3,0}P_{3,0} + B_{0,1}P_{0,1} + B_{1,1}P_{1,1} + B_{2,1}P_{2,1} + B_{3,1}P_{3,1}+ \\ &B_{0,2}P_{0,2} + B_{1,2}P_{1,2} + B_{2,2}P_{2,2} + B_{3,2}P_{3,2} + B_{0,3}P_{0,3} + B_{1,3}P_{1,3} + B_{2,3}P_{2,3} + B_{3,3}P_{3,3} \end{aligned} \\\\
\begin{aligned} &B_{0,0}P_{3,0} + B_{1,0}P_{4,0} + B_{2,0}P_{5,0} + B_{3,0}P_{6,0} + B_{0,1}P_{3,1} + B_{1,1}P_{4,1} + B_{2,1}P_{5,1} + B_{3,1}P_{6,1}+ \\ &B_{0,2}P_{3,2} + B_{1,2}P_{4,2} + B_{2,2}P_{5,2} + B_{3,2}P_{6,2} + B_{0,3}P_{3,3} + B_{1,3}P_{4,3} + B_{2,3}P_{5,3} + B_{3,3}P_{6,3} \end{aligned} \\\\
\begin{aligned} &B_{0,0}P_{0,3} + B_{1,0}P_{1,3} + B_{2,0}P_{2,3} + B_{3,0}P_{3,3} + B_{0,1}P_{0,4} + B_{1,1}P_{1,4} + B_{2,1}P_{2,4} + B_{3,1}P_{3,4}+ \\ &B_{0,2}P_{0,5} + B_{1,2}P_{1,5} + B_{2,2}P_{2,5} + B_{3,2}P_{3,5} + B_{0,3}P_{0,6} + B_{1,3}P_{1,6} + B_{2,3}P_{2,6} + B_{3,3}P_{3,6} \end{aligned} \\\\
\vdots \qquad\qquad \vdots \\\\
\begin{aligned} &B_{0,0}P_{n-3,m-3} + B_{1,0}P_{n-2,m-3} + B_{2,0}P_{n-1,m-3} + B_{3,0}P_{n,m-3} + B_{0,1}P_{n-3,m-2} + B_{1,1}P_{n-2,m-2}+ \\ &B_{2,1}P_{n-1,m-2} + B_{3,1}P_{n,m-2} + B_{0,2}P_{n-3,m-1} + B_{1,2}P_{n-2,m-1} + B_{2,2}P_{n-1,m-1} + B_{3,2}P_{n,m-1}+ \\ &B_{0,3}P_{n-3,m} + B_{1,3}P_{n-2,m} + B_{2,3}P_{n-1,m} + B_{3,3}P_{n,m} \end{aligned}
\end{bmatrix}
$$
(2)

In this equation, $B_{r,s} = B_r(u)B_s(v)$ and the control points $P_{i,j}$ should be computed to determine the free-form surface model. Thus, the control points are determined by the expression $P_{i,j} = z_{i,j}\, w_{i,j}$, where the weights $w_{i,j}$ are in the interval $[0.7, 1.3]$. Additionally, the initial Bezier surface Equation (1) is determined by means of $w_{i,j} = 1$, which provides the control points $P_{i,j} = z_{i,j}$. Additionally, the surface points are represented by the coordinates $(x_{i,j}, y_{i,j}, z_{i,j})$. Thus, the values (u, v) are computed by the expressions

$u = (x_{3+p*3,q} - x_{0+p*3,q})/3$ and $v = (y_{p,3+q*3} - y_{p,0+q*3})/3$, respectively. Based on these statements, the control points $P_{i,j}$ are determined by means of the weights $w_{i,j}$. In this way, a multi-objective optimization through a genetic algorithm is implemented to determine the weights $w_{i,j}$. To do so, the genetic algorithm is implemented as follows.

The first step is to compute the solution space and the initial population. Typically, the multi-objective optimization based on a genetic algorithm determines the solution space through the Pareto solution method [31], where the solution space is determined from the edge between the minimum of the objective functions, which is called Pareto front [32]. However, it is necessary to establish additional criteria to determine the best solution from the line or curve of the Pareto front. This leads to implementing additional procedures in the genetic algorithm. Additionally, the multi-objective optimization based on intelligent algorithms determined the solution space through the maximum and minimum [33]. Therefore, the proposed technique determines the solution space through the initial Bezier surface, which provides the maximum or minimum of each control $P_{i,j}$. In this way, the initial Bezier surface is computed via Equation (1) by employing $w_{i,j} = 1_j$. Thus, if $S_{p,q}(u,v)$ is over the surface point $z_{i,j}$, the maximum is defined as $z_{i,j}*1.0$ and the minimum is defined by $z_{i,j}*0.7$, which moves the Bezier surface under the $z_{i,j}$. On the other hand, if $S_{p,q}(u,v)$ is under the surface point $z_{i,j}$, the minimum is established as $z_{i,j}*1.0$ and the maximum is defined by $z_{i,j}*1.3$, which moves the Bezier surface over the $z_{i,j}$. Thus, the solution space is generated. In this way, the initial Bezier surface provides the solution space from the surface data to perform the multi-objective optimization. Then, the initial population is generated from the solution space. To carry it out, four values are randomly computed from the solution space for each weight. These values represent the parents $(\mathcal{P}_{1,k}, \mathcal{P}_{2,k})$, $(\mathcal{P}_{3,k}, \mathcal{P}_{4,k})$, where the k-index indicates the generation number. With these parents, the initial population is completed.

The second step is to perform the crossover to create the current children. The crossover performs exploration and exploitation to create two children inside the parents and two children outside the parents [34]. In this way, the children $(C_{1+4*t,k}, C_{2+4*t,k}, C_{3+4*t,k}, C_{4+4*t,k})$ are created through the parents $(\mathcal{P}_{1+2*t,k}, \mathcal{P}_{2+2*t,k})$ for $t = 0$ and $t = 1$. Thus, the children are computed by the expressions

$$C_{1+4*t,k} = \begin{cases} \mathcal{P}_{1+2*t,k} - 0.5\beta\left|\mathcal{P}_{1+2*t,k} - \text{minimum}\right|, & \text{if } \mathcal{P}_{1+2*t,k} < \mathcal{P}_{2+2*t,k} \\ \\ \mathcal{P}_{2+2*t,k} - 0.5\beta\left|\mathcal{P}_{2+2*t,k} - \text{minimum}\right|, & \text{if } \mathcal{P}_{2+2*t,k} < \mathcal{P}_{1+2*t,k} \end{cases} \tag{3}$$

$$C_{2+4*t,k} = 0.5\left[(\mathcal{P}_{1+2*t,k} + \mathcal{P}_{2+2*t,k}) - \beta\left|\mathcal{P}_{1+2*t,k} - \mathcal{P}_{2+2*t,k}\right|\right], \tag{4}$$

$$C_{3+4*t,k} = 0.5\left[(\mathcal{P}_{1+2*t,k} + \mathcal{P}_{2+2*t,k}) + \beta\left|\mathcal{P}_{1+2*t,k} - \mathcal{P}_{2+2*t,k}\right|\right], \tag{5}$$

$$C_{4+4*t,k} = \begin{cases} \mathcal{P}_{2+2*t,k} + 0.5\beta\left|\text{maximum} - \mathcal{P}_{2+2*t,k}\right|, & \text{if } \mathcal{P}_{1+2*t,k} < \mathcal{P}_{2+2*t,k} \\ \\ \mathcal{P}_{1+2*t,k} + 0.5\beta\left|\text{maximum} - \mathcal{P}_{1+2*t,k}\right|, & \text{if } \mathcal{P}_{2+2*t,k} < \mathcal{P}_{1+2*t,k} \end{cases} \tag{6}$$

In these equations, the probability distribution β is determined through the spread factor α, which is randomly computed in the interval $[0, 1]$. Thus, the probability distribution is $\beta = (2\alpha)^{1/2}$ if $\alpha > 0.5$; otherwise, $\beta = [2(1 - \alpha)]^{1/2}$. In this way, the children inside the parents are computed by Equations (4) and (5), and the children outside the parents are computed by Equation (3) through Equation (6). Thus, the children $(C_{1,k}, C_{2,k}, C_{3,k}, C_{4,k})$ are computed by Equation (3) through Equation (6) through the parents $(\mathcal{P}_{1,k}, \mathcal{P}_{2,k})$ and $t = 0$. Then, the children $(C_{5,k}, C_{6,k}, C_{7,k}, C_{8,k})$ are computed by Equation (3) through Equation (6) through the parents $(\mathcal{P}_{3,k}, \mathcal{P}_{4,k})$ and $t = 1$. Thus, the current children are obtained. To assemble the Bezier surfaces $S_{p,q}(u,v)$ with G^1 continuity, the boundary control points must be collinear [35]. Therefore, the boundary control points are computed by the expressions $P_{3+3*p,j} = (P_{3+3*p-1,j} + P_{3+3*p+1,j})/2$ and $P_{i,3+3*q} = (P_{i,3+3*q-1} + P_{i,3+3*q+1})/2$.

The third step is to evaluate the fitness through a multi-objective function, which is deduced by the Bezier surfaces $S_{0,0}(u,v)$, $S_{1,0}(u,v)$, ..., $S_{n/3,m/3}(u,v)$. Thus, the multi-objective function is defined by the expression

$$
\begin{bmatrix} F_{0,0} \\[2ex] F_{1,0} \\[2ex] F_{0,1} \\[2ex] \vdots \\[2ex] F_{n/3,m/3} \end{bmatrix} = \begin{bmatrix} \min\left\{ \frac{1}{16}\sqrt{\sum_{s=0}^{s=3}\sum_{r=0}^{r=3}\left[S_{0,0}(u,v)-z_{i,j}\right]^2} \right\} \\[3ex] \min\left\{ \frac{1}{16}\sqrt{\sum_{s=0}^{s=3}\sum_{r=0}^{r=3}\left[S_{1,0}(u,v)-z_{i,j}\right]^2} \right\} \\[3ex] \min\left\{ \frac{1}{16}\sqrt{\sum_{s=0}^{s=3}\sum_{r=0}^{r=3}\left[S_{0,1}(u,v)-z_{i,j}\right]^2} \right\} \\[3ex] \vdots \\[3ex] \min\left\{ \frac{1}{16}\sqrt{\sum_{i=0}^{i=3}\sum_{j=0}^{j=3}\left[S_{n/3,m/3}(u,v)-z_{i,j}\right]^2} \right\} \end{bmatrix}. \tag{7}
$$

From this multi-objective function, the fitness is computed by the expression $fitness = (F_{0,0} + F_{1,0} + F_{0,1} + \ldots + F_{n/3,m/3})/[(n/3)(m/3)]$.

The fourth step is to select the parents of the next generation through the best parents and children. Thus, the parents $\mathcal{P}_{1,k+1}$ and $\mathcal{P}_{3,k+1}$ are selected from the parents $(\mathcal{P}_{1,k}, \mathcal{P}_{2,k})$ and $(\mathcal{P}_{3,k}, \mathcal{P}_{4,k})$, respectively. Then, the parents $\mathcal{P}_{2,k+1}$ and $\mathcal{P}_{4,k+1}$ are selected from the children $(C_{1,k}, C_{2,k}, C_{3,k}, C_{4,k})$ and $(C_{5,k}, C_{6,k}, C_{7,k}, C_{8,k})$, respectively.

The fifth step is to perform the mutation. This procedure leads to avoiding trapping in a local minimum. To carry it out, a new parent is randomly generated from the search space. Then, the new parent replaces the worst parent, which is selected through the fitness Equation (7). Thus, if the new parent improves the fitness, the worst parent is replaced by the new parent. Otherwise, the worst parent is not replaced. Additionally, one weight is mutated from a parent, which is selected in random form. To do so, a new weight is randomly generated from the search space, and it is replaced by the selected parent to compute the fitness Equation (7). If the new weight improves the fitness, the mutation is carried out; if not, the weight is not mutated. Thus, the mutation is accomplished to determine the parents of the $(k + 1)$ generation. Then, the children of the $(k + 1)$ generation are generated through the crossover by computing Equation (3) through Equation (6). From this step, the population of the $(k + 1)$ is completed. The steps to determine the $(k + 1)$ generation are repeated until the multi-objective function in Equation (7) is minimized.

To elucidate the multi-objective optimization, a free-form surface model was used by employing the contour data shown in Figure 2a. This procedure was performed through the flowchart shown in Figure 2b, which describes the steps to perform the free-form surface modeling via multi-objective optimization. In this way, the first step was carried out to determine the initial population. To do so, the initial Bezier surface Equation (1) was computed to define the solution space through the maximum and minimum of each control point $P_{i,j}$. In this case, control point $P_{0,0}$ was provided by the initial Bezier surface, and the data $(P_{3,0}, P_{3,1}, P_{3,2}, P_{3,3}, P_{0,3}, P_{1,3}, P_{2,3})$ were computed through the expressions $P_{3+3*p,j} = (P_{3+3*p-1,j} + P_{3+3*p+1,j})/2$ and $P_{i,3+3*q} = (P_{i,3+3*q-1} + P_{i,3+3*q+1})/2$ to provide continuity G^1. Thus, the first parents were computed from the search space. The data of the initial population of the surface $S_{0,0}(u,v)$ are shown in Table 1. In this table, the first column indicates the control points, the second column indicates the generation number, and the parents $(\mathcal{P}_{1,1}, \mathcal{P}_{2,1}, \mathcal{P}_{3,1}, \mathcal{P}_{4,1})$ are indicated in the third to sixth column. Then, the second step was performed to compute the current children through the crossover. To carry it out, Equation (3) through Equation (6) were computed by employing the parents $(\mathcal{P}_{1+2*t,k}, \mathcal{P}_{2+2*t,k})$ and $t = 0$ to generate the children $(C_{1,k}, C_{2,k}, C_{3,k}, C_{4,k})$. In the same way, Equation (3) through Equation (6) were computed by employing the parents $(\mathcal{P}_{1+2*t,k},$

$\mathcal{P}_{2+2^*t,k}$) and $t = 1$ to determine ($C_{5,k}$, $C_{6,k}$, $C_{7,k}$, $C_{8,k}$). The children data are in the seventh to fourteenth column of Table 1. Next, the third step was performed to evaluate the fitness by means of the multi-objective function, which was deduced through the Bezier surfaces $S_{0,0}(u,v)$, $S_{1,0}(u,v)$, ... , $S_{1,1}(u,v)$. This procedure indicated that the initial population produced a low error. Then, the fourth step was carried out to select the parents of the next generation from the best current parents and children. In this way, the parents $\mathcal{P}_{1,k+1}$ and $\mathcal{P}_{3,k+1}$ were selected from the pairs ($\mathcal{P}_{1,k}$, $\mathcal{P}_{2,k}$) and ($\mathcal{P}_{3,k}$, $\mathcal{P}_{4,k}$), respectively. Additionally, the parents $\mathcal{P}_{2,k+1}$ and $\mathcal{P}_{4,k+1}$ were selected from the children ($C_{1,k}$, $C_{2,k}$, $C_{3,k}$, $C_{4,k}$) and ($C_{5,k}$, $C_{6,k}$, $C_{7,k}$, $C_{8,k}$), respectively. In this case, $\mathcal{P}_{1,2} = \mathcal{P}_{2,1}$, $\mathcal{P}_{3,2} = \mathcal{P}_{4,1}$, $\mathcal{P}_{2,2} = C_{1,1}$, and $\mathcal{P}_{4,2} = C_{5,1}$. Then, the fifth step was performed to mutate the parent $\mathcal{P}_{1,2}$, which was selected as the worst parent. Additionally, a new parent was randomly generated from the solution space to compute the fitness (7). In this case, the fitness was not improved and the parent was not mutated. Then, the parent $\mathcal{P}_{1,2}$ was chosen in random form to mutate the weight $w_{1,2}$, which was randomly selected. Next, a new weight was replaced in the parent $\mathcal{P}_{1,2}$, and the fitness was computed via Equation (7). In this case, the fitness was improved; therefore, the weight $w_{1,2}$ was mutated. Then, the second step was carried out by computing Equation (3) through Equation (6) to create the children of the $(k + 1)$ generation. Further, the fitness of these children was computed via Equation (7). Additionally, the control points ($P_{3,0}$, $P_{3,1}$, $P_{3,2}$, $P_{3,3}$, $P_{0,3}$, $P_{1,3}$, $P_{2,3}$) were determined by the expressions $P_{3+3^*p,j} = (P_{3+3^*p-1,j} + P_{3+3^*p+1,j})/2$ and $P_{i,3+3^*q} = (P_{i,3+3^*q-1} + P_{i,3+3^*q+1})/2$ to provide continuity G^1.

The second-generation population is indicated in the tenth to twenty-fourth row of Table 1. In this way, the steps to compute the $(k+1)$ generation were repeated until the multi-objective function in Equation (7) was minimized. The optimal control points are shown in the fifteenth column of Table 1. These control points define the free-form surface model, which generates the surface shown in Figure 2c. Thus, the free-form surface modeling was performed through multi-objective optimization. Additionally, this method can be performed to construct a free-form surface model for a non-rectangular surface. In this case, the contour in the x-axis and y-axis should be represented through the Bezier functions $X_{p,q}(u,v)$ and $Y_{p,q}(u,v)$, respectively. These functions are deduced via Equation (1) by the expressions $X_{p,q}(u,v) = B_0(u)B_0(v)x_{0,0} + B_1(u)B_0(v)x_{1,0} + \ldots + B_3(u)B_3(v)x_{3,3}$ and $Y_{p,q}(u,v) = B_0(u)B_0(v)y_{0,0} + B_1(u)B_0(v)y_{1,0} + \ldots + B_3(u)B_3(v)y_{3,3}$. Thus, the free-form surface model for a non-rectangular surface is defined by $\{X_{p,q}(u,v), X_{p,q}(u,v), S_{p,q}(u,v)\}$.

The efficiency of the multi-objective optimization is elucidated through the parameters in the genetic algorithm. The parameters include the population size, number of generations, crossover probability and mutation probability, and optimal gap. The population size indicates the number of chromosomes in one generation. In this case, each control point included an initial population of twelve chromosomes. However, each surface function $S_{p,q}(u,v)$ included eight control points. Therefore, one generation included a population 384 chromosomes. The crossover probability indicates how often the crossover is performed. The crossover is made to find better chromosomes. However, it is good to leave some of the old population for the next generation. The probability of crossover is determined via fitness [36]. When the fitness average is improved, the crossover is carried out. This procedure avoids the loss of candidates to achieve the convergence. In this way, the multi-objective optimization is performed several times for the same free-form surface model. The result of the probability of crossover was in the interval from 0.18 to 0.56. The mutation probability indicates how often the chromosome can be mutated. The mutation probability is determined via fitness. In the genetic algorithm, if the fitness is improved, the mutation is carried out; if not, the parameter is not mutated. In this way, the multi-objective optimization is performed several times on the same free-form surface model. The result of the probability of mutation was in the interval from 0.26 to 0.61. The number of generations indicates the number of iterations to obtain the optimal control points. In this case, 92 generations were performed to optimize the parameters. The flat surface modeling via multi-objective optimization is described in Section 2.2. The optimal gap was computed through the relative error, and the result was 0.38%.

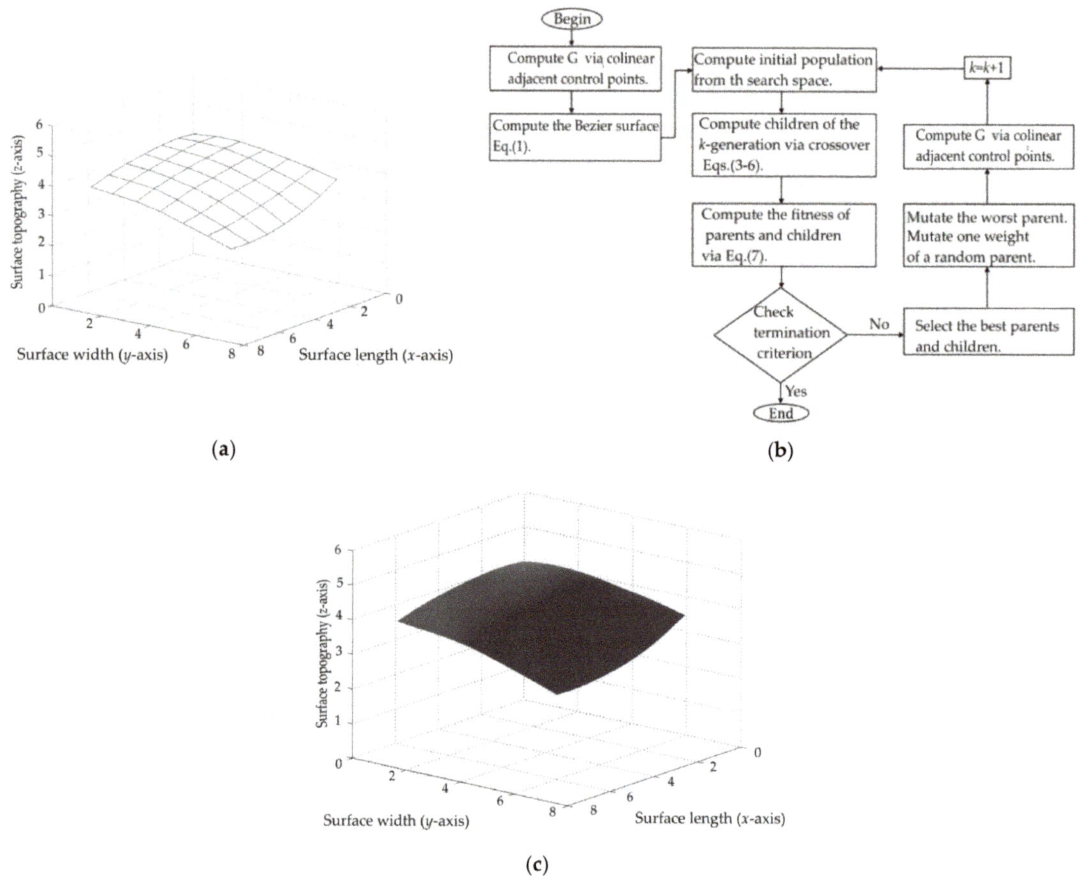

Figure 2. (**a**) Contour data to construct free-form surface model. (**b**) Flowchart to perform multi-objective optimization through a genetic algorithm. (**c**) Surface generated by the free-form surface model Equation (1).

Table 1. Control points generated in the first and second generation.

$P_{i,j}$	k	$\mathcal{P}_1$	$\mathcal{P}_2$	$\mathcal{P}_3$	$\mathcal{P}_4$	C_1	C_2	C_3	C_4	C_5	C_6	C_7	C_8	
$P_{1,0}$	1	4.781	4.766	4.375	4.740	4.668	4.768	4.775	4.733	4.364	4.451	4.672	4.199	
$P_{2,0}$	1	4.329	4.645	4.238	4.064	4.254	4.456	4.527	4.605	4.063	4.071	4.153	4.216	
$P_{0,1}$	1	4.753	4.144	4.550	4.698	4.131	4.175	4.514	4.737	4.458	4.623	4.695	4.501	
$P_{1,1}$	1	4.396	4.643	4.771	4.394	4.266	4.479	4.636	4.601	4.265	4.428	4.755	4.716	
$P_{2,1}$	1	4.163	4.487	4.549	4.533	4.162	4.184	4.486	4.419	4.382	4.537	4.546	4.468	
$P_{0,2}$	1	4.724	4.117	4.736	4.094	4.085	4.375	4.439	4.660	4.092	4.097	4.570	4.663	
$P_{1,2}$	1	4.318	4.543	4.657	4.484	4.215	4.382	4.450	4.514	4.457	4.512	4.643	4.633	
$P_{2,2}$	1	4.346	4.118	4.525	4.508	4.084	4.210	4.311	4.259	4.433	4.509	4.519	4.449	
fitness		0.258	0.179	0.193	0.166	0.106	0.188	0.256	0.269	0.085	0.130	0.222	0.159	
$P_{1,0}$	2	4.766	4.668	4.740	4.364	4.380	4.670	4.748	4.754	4.221	4.447	4.625	4.685	4.135
$P_{2,0}$	2	4.645	4.254	4.064	4.063	4.234	4.320	4.535	4.569	4.058	4.063	4.064	3.880	4.074
$P_{0,1}$	2	4.144	4.131	4.698	4.458	4.104	4.133	4.143	3.803	4.303	4.459	4.587	4.683	4.154

Table 1. *Cont.*

$P_{1,1}$	2	4.643	4.266	4.394	4.265	4.186	4.440	4.464	4.605	4.184	4.327	4.371	4.283	4.195
$P_{2,1}$	2	4.487	4.162	4.533	4.382	4.130	4.314	4.454	4.348	4.359	4.420	4.481	4.473	3.992
$P_{0,2}$	2	4.117	4.085	4.094	4.092	4.067	4.096	4.114	4.001	4.073	4.092	4.093	3.786	4.112
$P_{1,2}$	2	4.059	4.215	4.484	4.457	4.028	4.071	4.178	3.935	4.395	4.457	4.481	4.421	4.163
$P_{2,2}$	2	4.118	4.084	4.507	4.433	3.998	4.088	4.115	3.900	4.232	4.452	4.487	4.432	3.981
fitness		0.179	0.106	0.166	0.085	0.062	0.120	0.155	0.155	0.054	0.097	0.140	0.112	0.0126

2.2. Micro-Scale Flat Surface Modeling via Multi-Objective Optimization

The micro-scale flat surface modeling is performed through the multi-objective optimization and surface points, which are contoured via micro laser line projection. Thus, a flat surface model is constructed through the surface coordinates. The flat surface points are shown in Figure 3. The surface points are represented by the coordinates $(x_{i,j}, y_{i,j}, z_{i,j})$, where the sub-indices (i, j) are defined in the x-axis and y-axis, respectively.

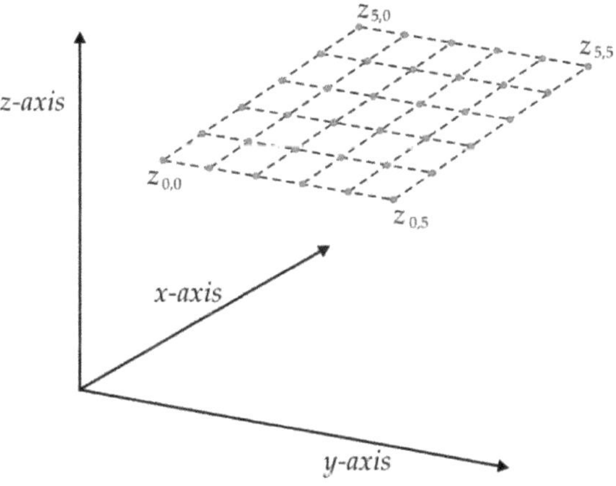

Figure 3. Surface coordinates to perform flat surface modeling.

Thus, the flat surface provides the coordinates $(z_{0,0}, z_{1,0}, \dots, z_{n,m})$ in the z-axis, where the sub-indices (n, m) represent the number of coordinates in the x-axis and y-axis, respectively. Thus, the flat surface model is deduced by the next plane expression

$$z_{i,j} = ax_{i,j} + by_{i,j} + c. \tag{8}$$

Additionally, the derivatives in the x-axis and y-axis are defined through the surface coordinates by means of the next expressions

$$Gx = \frac{z_{i+\delta,j} - z_{i,j}}{x_{i+\delta,j} - x_{i,j}} = \frac{(ax_{i+\delta,j} + by_{i+\delta,j}) - (ax_{i,j} + by_{i,j})}{x_{i+\delta,j} - x_{i,j}}, \tag{9}$$

$$Gy = \frac{z_{i,j+\Delta} - z_{i,j}}{y_{i,j+\Delta} - y_{i,j}} = \frac{(ax_{i,j+\Delta} + by_{i,j+\Delta}) - (ax_{i,j} + by_{i,j})}{y_{i,j+\Delta} - y_{i,j}}. \tag{10}$$

In these derivatives, δ and Δ represent an increment in the x-axis and y-axis, respectively. Thus, the flat surface model is established by computing the constants (a, b, c). In this way, the constants (a, b, c) are optimized to establish the flat surface model. To carry it out, a multi-objective optimization is performed via Equations (8)–(10) and the surface coordinates $(x_{i,j}, y_{i,j}, z_{i,j})$. Thus, the genetic algorithm determines the constants (a, b, c) as follows.

The first step is to determine the initial population by means of the surface contour coordinates. To carry it out, the initial constants (a, b, c) are computed by means of the next expressions

$$a = \frac{(z_{i+\delta,j} - z_{i,j})(y_{i,j+\Delta} + y_{i,j}) - (z_{i,j+\Delta} - z_{i,j})(y_{i+\delta,j} - y_{i,j})}{(x_{i+\delta,j} - x_{i,j})(y_{i,j+\Delta} + y_{i,j}) - (x_{i,j+\Delta} - x_{i,j})(y_{i+\delta,j} - y_{i,j})}, \tag{11}$$

$$b = \frac{(z_{i+\delta,j} - z_{i,j})(x_{i,j+\Delta} - x_{i,j}) - (z_{i,j+\Delta} - z_{i,j})(x_{i+\delta,j} - x_{i,j})}{(y_{i+\delta,j} - y_{i,j})(x_{i,j+\Delta} - x_{i,j}) - (y_{i,j+\Delta} + y_{i,j})(x_{i+\delta,j} - x_{i,j})}, \tag{12}$$

$$c = z_{i,j} - a x_{i,j} - b y_{i,j}. \tag{13}$$

In this way, four parents $(\mathcal{P}_{1,k}, \mathcal{P}_{2,k}, \mathcal{P}_{3,k}, \mathcal{P}_{4,k})$ are obtained by computing Equation (11) through Equation (13) for each constant. The sub-index k indicates the number of generations. Thus, the parent $\mathcal{P}_{1,1}$ is determined by computing Equation (11) through Equation (13) by employing ($i = 0, j = 0, \delta = n/2, \Delta = m/2$). In the same way, the parent $\mathcal{P}_{2,1}$ is computed by employing ($i = n/2, j = 0, \delta = n, \Delta = m/2$), $\mathcal{P}_{3,1}$ is computed by employing ($i = 0, j = m/2, \delta = n/2, \Delta = m$), and $\mathcal{P}_{4,1}$ is computed by employing ($i = n/2, j = m/2, \delta = n, \Delta = m$). Thus, four constants (a, b, c) are obtained, and they represent the initial population. Additionally, the mean and the standard deviation are computed from the four constants, and they are defined as (am, bm, cm) and (as, bs, cs), respectively. Then, the solution space is determined through the maximum and minimum of each constant. Thus, the terms $(am - 2*as)$ and $(am + 2*as)$ are computed to determine the maximum and minimum of a. Additionally, the terms $(bm - 2*as)$ and $(bm + 2*bs)$ are computed to determine the maximum and minimum of b. Further, the terms $(cm - 2*cs)$ and $(cm + 2*cs)$ are computed to determine the maximum and minimum of c. Thus, the solution space is defined, and the initial population is completed.

The second step is to generate the current children through the crossover. Thus, the parents $(\mathcal{P}_{1,k}, \mathcal{P}_{2,k})$ and $t = 0$ are replaced in Equation (3) through Equation (6) to compute the children $(C_{1,k}, C_{2,k}, C_{3,k}, C_{4,k})$. Additionally, the parents $(\mathcal{P}_{3,k}, \mathcal{P}_{4,k})$ and $t = 1$ are replaced in Equation (3) through Equation (6) to compute the children $(C_{5,k}, C_{6,k}, C_{7,k}, C_{8,k})$.

The third step is to determine the fitness through a multi-objective function, which is deduced by means of Equation (8) through Equation (10) and the surface coordinates $(x_{i,j}, y_{i,j}, z_{i,j})$. The multi-objective function is defined by the expression

$$\begin{bmatrix} F_1 \\ F_2 \\ F_3 \end{bmatrix} = \begin{bmatrix} \min\left\{ \sqrt{\sum_{i=1}^{i=n}\sum_{j=1}^{r=m} [z_{i,j} - (a x_{i,j} + b y_{i,j} - c)]^2} \right\} \\ \min\left\{ \sqrt{\sum_{i=1}^{i=n-1}\sum_{j=1}^{j=m} [(z_{i+\delta,j} - z_{i,j}) - a(x_{i+\delta,j} - x_{i,j}) - b(y_{i+\delta,j} - y_{i,j})]^2} \right\} \\ \min\left\{ \sqrt{\sum_{i=1}^{i=n}\sum_{j=1}^{j=m-1} [(z_{i,j+\Delta} - z_{i,j}) - a(x_{i,j+\Delta} - x_{i,j}) - b(y_{i,j+\Delta} - y_{i,j})]^2} \right\} \end{bmatrix}. \tag{14}$$

The fourth step is to select the parents of the next generation from the best current parents and children. Thus, the parents $\mathcal{P}_{1,k+1}$ and $\mathcal{P}_{3,k+1}$ are selected from the pairs $(\mathcal{P}_{1,k}, \mathcal{P}_{2,k})$ and $(\mathcal{P}_{3,k}, \mathcal{P}_{4,k})$, respectively. Then, the parents $\mathcal{P}_{2,k+1}$ and $\mathcal{P}_{4,k+1}$ are selected from the children $(C_{1,k}, C_{2,k}, C_{3,k}, C_{4,k})$ and $(C_{5,k}, C_{6,k}, C_{7,k}, C_{8,k})$, respectively.

The fifth step is to perform the mutation, which leads to avoiding trapping in a local minimum. To carry it out, a new parent is randomly generated from the search space. Then, the new parent replaces the worst parent, which is selected by means of the fitness Equation (14). Thus, if the new parent improves the fitness, the worst parent is replaced by the new parent. Otherwise, the worst parent is not mutated. Additionally, one constant is mutated from a parent, which is selected in random form. To do so, a new constant is randomly generated from the solution space, and it is replaced in the selected parent to compute the fitness via Equation (14). If the new constant improves the fitness, the mutation is carried out; if not, the constant is not mutated. Thus, the mutation is accomplished to determine the parents of the $(k + 1)$ generation. Then, the children of the $(k + 1)$ generation are generated through the crossover by computing Equation (3) through Equation (6). Additionally, the fitness of these children is computed via Equation (14). From this step, the population of the $(k + 1)$ generation is completed. Then, the steps to determine the $(k + 1)$ generation are repeated until the multi-objective function in Equation (14) is minimized. After that, the optimal constants (a, b, c) are replaced in Equation (8) to obtain the flat surface model. Thus, the micro-scale flat surface model is generated by the multi-objective optimization through a genetic algorithm.

2.3. Surface Contouring at Micro-Scale via Micro Laser Line Scanning

The microscope vision system to perform micro-scale surface contouring is shown in Figure 4a. This arrangement consists of an optical microscope, which includes a laser CCD camera, a diode, and a computer. This microscope system is mounted on a slider device to perform the micro laser scanning. In this arrangement, the surface plane is located on the x-axis and y-axis, and the surface contour is parallel to the z-axis. The lateral view of the microscope arrangement in the x-axis is described by the optical geometry shown in Figure 4b, where a 38 µm laser line is projected perpendicularly on the target surface and the microscope is aligned at an angle. The symbol θ depicts the angle between the optical axis and the laser line. The distance between the surface and the objective lens is defined by d. The objective focal length and the objective focus are represented by L_1 and F_1, respectively. The distance between the ocular lens and the intermediate image plane is represented by L. The ocular focal length and the ocular focus are represented L_2 and F_2, respectively. The lateral view of the microscope arrangement in the y-axis is described by the optical geometry shown in Figure 4c. The laser line coordinates in the image plane are represented by $(x_{i,j}, y_{i,j})$ in the x-axis and y-axis, respectively. The image center is represented by the coordinates (x_c, y_c), and the pixel size is represented by η. The topography coordinates $y_{i,j}$ and $z_{i,j}$ are deduced from the geometry shown in Figure 4b,c by means of the expressions

$$z_{i,j} = \frac{\eta(x_c - x_{i,j})F_1F_2}{(L_1 - F_1)(L_2 - F_2)\sin\theta} + O, \tag{15}$$

$$y_{i,j} = \eta y_c - \frac{\eta(y_c - y_{i,j})F_1F_2}{(L_1 - F_1)(L_2 - F_2)}. \tag{16}$$

The surface coordinate $x_{i,j}$ is collected from the position where the laser line is projected in the x-axis. The slider device provides the coordinate in the x-axis. Thus, the topography coordinates are computed through the parameters $(x_c, y_c, \eta, \theta, L_1, F_1, L_2, F_2)$. These vision parameters are determined by a genetic algorithm via Equations (15) and (16) and the topography coordinates. The genetic algorithm computes the vision as follows.

The first step is to compute the initial population from the maximum and minimum of each parameter. In this way, the maximum and minimum of the constants (x_c, y_c, η) are deduced from the image size. However, the maximum and minimum of the variables $(\theta, L_1, F_1, L_2, F_2)$ are defined through the microscope geometry. The minimum F_2 is deduced through the ocular lens diameter, and the maximum F_2 is two times the minimum F_2. The minimum L_2 is equal to the minimum F_2, and the maximum L_2 is two times the minimum L_2. Additionally, the minimum F_1 is deduced through the objective lens diameter, and the maximum F_1 is two times the minimum F_1. The minimum L_1 is equal to the minimum F_1, and the maximum L_1 is two times the minimum L_1. The minimum and minimum θ were established as $14°$ and $40°$, respectively. Thus, the solution space is completed. Then, four parents $(\mathcal{P}_{1,k}, \mathcal{P}_{2,k}, \mathcal{P}_{3,k}, \mathcal{P}_{4,k})$ are computed from the solution space in random form.

In this way, the four vision parameters $(x_c, y_c, \eta, \theta, L_1, f_1, L_2, f_2)$ are established as the initial population.

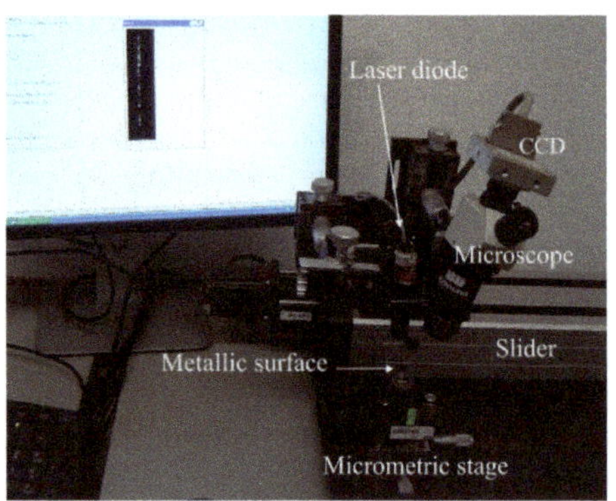

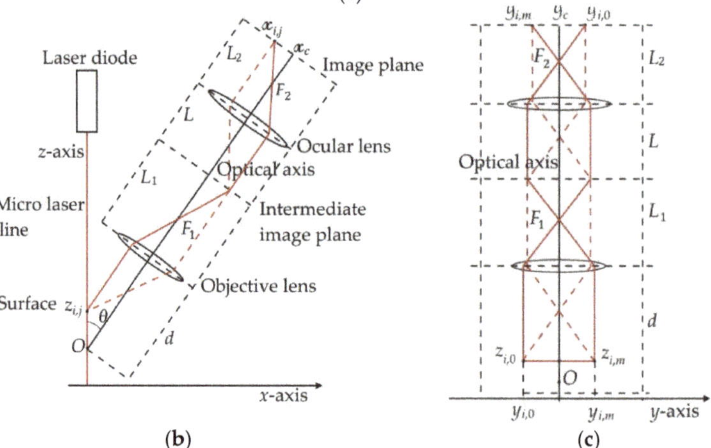

Figure 4. (**a**) Microscope vision system to compute micro-scale flat and free-form topography [37]. (**b**) Lateral geometry of the microscope system in x-axis. (**c**) Geometry of the microscope system in y-axis.

The second step is to perform the crossover to generate the current children. To carry it out, the children $(C_{1+4*t,k}, C_{2+4*t,k}, C_{3+4*t,k}, C_{4+4*t,k})$ are computed through the parents $(\mathcal{P}_{1+2*t,k}, \mathcal{P}_{2+2*t,k})$ for $t = 0$ and $t = 1$. Thus, the parents $(\mathcal{P}_{1,k}, \mathcal{P}_{2,k})$ and $t = 0$ are replaced in Equation (3) through Equation (6) to compute the children $(C_{1,k}, C_{2,k}, C_{3,k}, C_{4,k})$. In the same way, the parents $(\mathcal{P}_{3,k}, \mathcal{P}_{4,k})$ and $q = 1$ are substituted in Equation (3) through Equation (6) to compute the children $(C_{5,k}, C_{6,k}, C_{7,k}, C_{8,k})$.

The third step is to evaluate the fitness through an objective function, which is deduced from the vision parameters by means of the expressions

$$FO_1 = \min\left\{\frac{1}{mxn}\sum_{i=}^{n}\sum_{j=0}^{m}\left[(z_{i,j} - z_{i,m}) - \frac{\eta(x_c - x_{i,j})F_1F_2}{(L_1 - F_1)(L_2 - F_2)\sin\theta} + \frac{\eta(x_c - x_{i,m})F_1F_2}{(L_1 - F_1)(L_2 - F_2)\sin\theta}\right]^2\right\}, \tag{17}$$

$$FO_2 = \min\left\{\frac{1}{mxn}\sum_{i=}^{n}\sum_{j=0}^{m}\left[(y_{i,j} - y_{i,m}) + \frac{\eta(y_c - y_{i,j})F_1F_2}{(L_1 - F_1)(L_2 - F_2)} - \frac{\eta(y_c - y_{i,m})F_1F_2}{(L_1 - F_1)(L_2 - F_2)}\right]^2\right\}. \tag{18}$$

From these objective functions, the fitness is evaluated through the expression $FO = (FO_1 + FO_2)/2$, where the surface topography $(z_{i,j} - z_{i,m})$ and the surface width $(y_{i,j} - y_{i,m})$ are known.

The fourth step is to select the parents of the next generation from the best current parents and children. In this way, the parents $\mathcal{P}_{1,k+1}$ and $\mathcal{P}_{3,k+1}$ are chosen from the parents $(\mathcal{P}_{1,k}, \mathcal{P}_{2,k})$ and $(\mathcal{P}_{3,k}, \mathcal{P}_{4,k})$, respectively. Then, the parents $\mathcal{P}_{2,k+1}$ and $\mathcal{P}_{4,t+1}$ are chosen from the children $(C_{1,k}, C_{2,k}, C_{3,k}, C_{4,k})$ and $(C_{5,k}, C_{6,k}, C_{7,k}, C_{8,k})$, respectively.

The fifth step is to perform the mutation to avoid trapping in a local minimum. To carry it out, a new parent is randomly generated from the solution space. Then, the new parent replaces the worst parent, which is selected through the fitness Equations (17) and (18). Thus, if the new parent improves the fitness, the worst parent is replaced by the new parent. Otherwise, the worst parent is not replaced. Additionally, one vision parameter is mutated from a parent, which is selected in random form. To do so, a new vision parameter is randomly generated from the solution space, and it is replaced in the selected parent to compute the fitness by means of Equations (17) and (18). If the new vision parameter improves the fitness, the mutation is carried out; if not, the parameter is not mutated. Thus, the mutation is accomplished to determine the parents of the $(k + 1)$ generation. Then, the children of the $(k + 1)$ generation are generated through the crossover by computing Equation (3) through Equation (6). Additionally, the fitness of these children is computed by means of Equations (17) and (18). From this step, the population of the $(k + 1)$ generation is completed. The steps to determine the $(k + 1)$ generation are repeated until the objective function in Equations (17) and (18) is minimized. Then, the distance from zero to the point O is computed by the expression $z_{0,j} = \eta(x_{0,j} - x_c) F_1 F_2/(L_1 - F_1)(L_2 - F_2)\sin\theta$.

The laser line coordinates $(x_{i,j}, y_{i,j})$ are determined through the pixel gray-level. The position $x_{i,j}$ was determined through the gray-level maximum in the x-axis of the image [38]. In this way, the pixel gray-level is approximated to a Bezier curve in the x-direction by means of the expressions

$$x(u) = \sum_{i=0}^{N} C_i(1-u)^{N-i}u^i x_{i,j}, \ C_i = C_{i-1}(N+1-i)/i, \ C_0 = 1, \ \ 0 \le u \le 1. \quad (19)$$

$$I(u) = \sum_{i=0}^{N} C_i(1-u)^{N-i}u^i I_{i,j}, \ C_i = C_{i-1}(N+1-i)/i, \ C_0 = 1, \ \ 0 \le u \le 1. \quad (20)$$

In Equation (19), $x_{i,j}$ represents the laser line coordinates in the x-axis, and N represents the laser line width in pixels. The pixel gray-level is represented by $I_{i,j}$ in Equation (20), where the sub-index i indicates the pixel number in the x-axis and the sub-index j indicates the pixel number in the y-axis. In this way, $x_{i,j}$ is substituted in Equation (19) and $I_{i,j}$ is substituted Equation (20) to compute a concave curve $\{x(u), I(u)\}$ in the interval $0 \le u \le 1$. Thus, the second derivative $I''(u)$ is positive, and the gray-level maximum is computed through the derivative $I'(u) = 0$. The value u that provides $I'(u) = 0$ is computed through the bisection method. By substituting u in Equation (19), $x(u)$ is computed to determine the laser line position $x_{i,j} = x(u)$ in the x-axis. The laser line position $y_{i,j}$ is deduced from the row number in the y-axis. Additionally, the line edges $y_{i,0}$ and $y_{i,m}$ are determined by computing the first derivative in the y-axis. In this way, the coordinates $(x_{i,j}, y_{i,j})$ are replaced in Equation (15) and Equation (16) to compute the surface topography $(y_{i,j}, z_{i,j})$ at the micro-scale. To carry it out, the surface is scanned, and a laser line image is captured by the CCD. From this laser line image, the coordinates $(x_{i,j}, y_{i,j})$ are computed by means of Equations (19) and (20). Then, the coordinates $(x_{i,j}, y_{i,j})$ are substituted in Equations (15) and (16) to compute the surface topography $(y_{i,j}, z_{i,j})$. The slider device provides the coordinates $x_{i,j}$ in the x-axis. Thus, the micro-scale surface coordinates are determined.

3. Results of Micro-Scale Flat and Free-Form Surface Modeling

The micro-scale flat and free-form surface modeling was performed through the optical microscope system shown in Figure 4a. The free-form surface modeling at the micro-scale was implemented for the metallic object shown in Figure 5a. In this figure, the scale is indicated in millimeters in the *x*-axis. Additionally, Figure 5b shows the micro laser line projected on the object surface. In this way, the micro laser line scanned the object surface in the *x*-axis, and the laser line image was captured by the CCD camera. From the image, the line position coordinates $(x_{i,j}, y_{i,j})$ were computed by means of Equations (19) and (20), respectively. Then, the coordinate $x_{i,j}$ was substituted in Equation (15) and the coordinate $y_{i,j}$ was replaced in Equation (16) to compute the surface data $(z_{i,j}, y_{i,j})$. The slider device provided the surface data $x_{i,j}$.

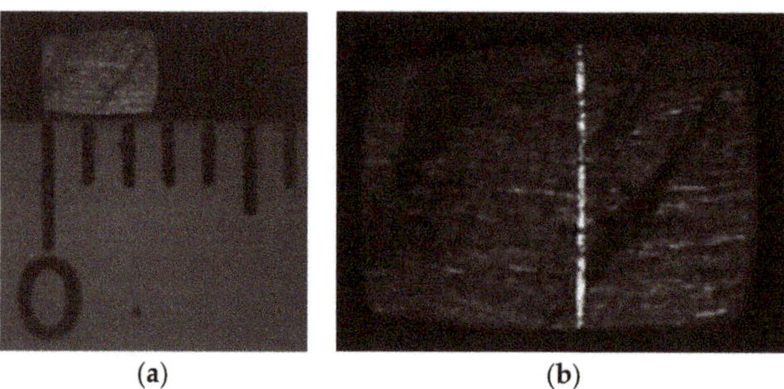

(a)　　　　　**(b)**

Figure 5. (**a**) Metallic object surface with scale in *mm* in *x*-axis. (**b**) Micro laser line projected on the object surface.

From the scanning, one hundred and sixty-eight images were captured to compute the object topography shown in Figure 6a. In this figure, the *x*-axis and *y*-axis are represented in millimeters, but the *z*-axis is represented in microns. The surface accuracy is computed through the relative error [39], which was calculated via reference data provided by a contact method. In this way, the surface accuracy was determined in terms of percentage by means of the expression

$$Error\% = \frac{100}{n \cdot m} \sum_{i=0}^{n} \sum_{j=0}^{m} \frac{\left| z_{i,j} - H_{i,j} \right|}{H_{i,j}},$$

(21)

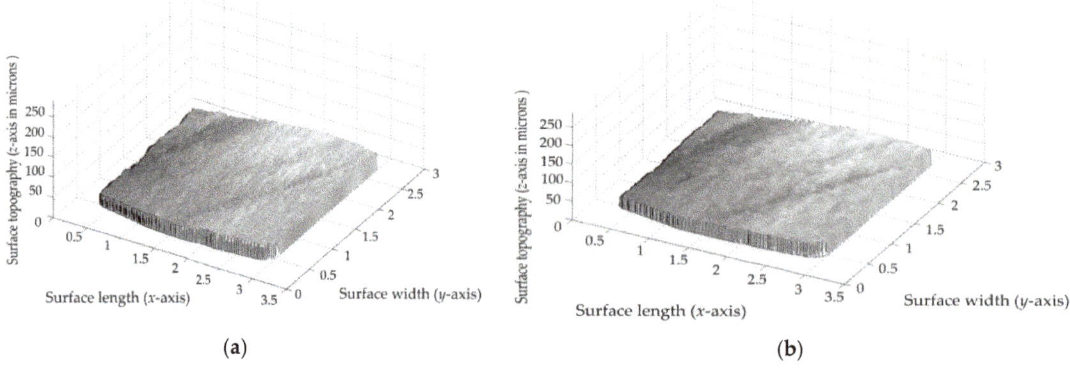

(a)　　　　　**(b)**

Figure 6. (**a**) Micro-scale topography recovered via micro laser line scanning. (**b**) Object surface computed by the free-form surface model generated via Equation (1).

In this equation, $z_{i,j}$ represents the micro-scale surface calculated by means of Equation (15), $H_{i,j}$ represents the surface reference, and $n \cdot m$ indicates the data number.

Thus, Equation (21) was computed by employing the surface data shown in Figure 6a, and the relative error was 0.785%. From the surface shown in Figure 6a, a free-form surface model was constructed through the multi-objective optimization as described in Section 2.1. The steps of this procedure are described as follows.

The first step was to determine the initial population from the solution space of each weight, where the minimum and maximum of the weights were in the interval [0.7, 1.3]. In this way, the genetic algorithm computed four parents in random form from the solution space for each weight of the surface $S_{p,q}(u,v)$. Then, the second step was to compute the current children through the crossover by replacing $t = 0$ and $t = 1$ in Equation (3) through Equation (6). The probability of crossover is deduced by means of the average fitness. Thus, the crossover is performed when the average fitness is enhanced. In this way, the loss of candidates is avoided to achieve the convergence. Then, the third step was to evaluate the fitness by substituting the weights of the parents and children in Equation (1) to compute Equation (7). Then, the fourth step was to select the parents of the next generation from the best current parents and children. Thus, the parents $\mathcal{P}_{1,k+1}$ and $\mathcal{P}_{3,k+1}$ were chosen from the pairs $(\mathcal{P}_{1,k}, \mathcal{P}_{2,k})$ and $(\mathcal{P}_{3,k}, \mathcal{P}_{4,k})$, respectively. In addition, the parents $\mathcal{P}_{2,t+1}$ and $\mathcal{P}_{4,t+1}$ were chosen from the children $(C_{1,k}, C_{2,k}, C_{3,k}, C_{4,k})$ and $(C_{5,k}, C_{6,k}, C_{7,k}, C_{8,k})$, respectively. Then, the fifth step was to mutate the worst parent, which was selected through the fitness. Thus, if the new parent enhanced the fitness, the worst parent was mutated. Otherwise, the worst parent was not mutated. In addition, one weight was mutated from a parent, which was selected in random form. To do so, a new weight was randomly generated from the solution space, and it was replaced in the selected parent to compute the fitness by means of Equation (7). If the new weight improved the fitness, the mutation was carried out; if not, the weight was not mutated. Then, the second step was carried out by computing Equation (3) through Equation (6) to create the children of the $(k + 1)$ generation. In addition, the fitness of these children was computed by means of Equation (7). From this step, the population of the $(k + 1)$ generation was completed. The steps to determine the $(k + 1)$ generation were repeated until the multi-objective function in Equation (7) was minimized.

The iterations to obtain the optimal weights determine the number of generations. In this case, 174 generations were computed to accomplish the free-form surface model. Thus, the optimal weights were substituted in Equation (1) to obtain the free-form surface model, which generated the object surface shown in Figure 6b. The fitting accuracy of the free-form surface model was determined through the relative error by the expression

$$Error\% = \frac{100}{n \cdot m} \sum_{i=0}^{n} \sum_{j=0}^{m} \frac{\left| M_{i,j} - z_{i,j} \right|}{z_{i,j}}, \tag{22}$$

In this equation, $M_{i,j}$ represents the micro-scale surface calculated through the free-form surface model Equation (1), $z_{i,j}$ represents the surface calculated by means of Equation (15), and $n \cdot m$ indicates the number of data. Additionally, the optimal gap of the genetic algorithm was determined by computing the relative error by means of Equation (22). From this step, the free-form surface model produced a relative error of 1.9731% with respect to the metallic surface shown in Figure 6a. The fitness variation indicated that the fitness decreased when the generation number increased.

The efficiency of the multi-objective optimization is described through the parameters in the genetic algorithm as follows. The population size is determined by the number of chromosomes in one generation. In this case, the free-surface model was established through the surface functions $S_{0,0}(u,v)$, $S_{1,0}(u,v)$, $S_{2,0}(u,v)$, $\ldots$, $S_{8,7}(u,v)$. Each surface function contained a population of 96 chromosomes. Therefore, one generation included a population of 6912 chromosomes. The crossover probability determines how often the crossover is performed. The probability of crossover is determined via fitness. Thus, the crossover is carried out when the average fitness of the parents is improved. In this way, the multi-objective optimization is performed several times at the same free-form surface model. The result of the probability of crossover was in the interval from 0.27 to

0.53. The mutation probability determines how often the chromosome can be mutated. The mutation probability is determined via fitness. Thus, if the new parent improves the fitness, the worst parent is mutated. In the same way, for the parameter mutation, if the new parameter improves the fitness, the parameter is mutated. Thus, the multi-objective optimization is performed several times on the same free-form surface model. The result of the probability of mutation was in the interval from 0.26 to 0.59. The number of generations indicates the iterations to obtain the optimal control points. In this case, 174 generations were performed to accomplish the free-form surface model. The optimal gap was computed via Equation (22), and the result was a relative error of 1.9731%. Based on these results, the genetic algorithm was examined. For instance, the optimal crossover probability and mutation probability were defined in the interval between 0.3 and 0.6 for surface modeling in recent optimization research [40]. Therefore, the proposed genetic algorithm provided good crossover probability and mutation probability. The population size is related to the convergence and the number of parameters [41]. In this case, the genetic algorithm provided 12 chromosomes for each parameter. These chromosomes produced results near the optimal solution and reduced iterations. Therefore, the algorithm provided a good population size and number of iterations. The optimal gap established good fitting of the surface model to the target surface.

The micro-scale flat surface modeling was carried out for the paper surface shown in Figure 7a. In this way, the paper surface was scanned in the x-axis to retrieve the coordinates $(x_{i,j}, y_{i,j})$ by computing Equations (19) and (20) from the laser line image. Then, the coordinate $x_{i,j}$ was replaced in Equation (15) and the coordinate $y_{i,j}$ was replaced in Equation (16) to calculate the surface data $(z_{i,j}, y_{i,j})$. The slider device provided the surface coordinates $x_{i,j}$. In this way, one hundred and sixty-four images were captured from the scanning to compute the object topography shown in Figure 7b. In this figure, the x-axis and y-axis are represented in millimeters, but the z-axis is represented in microns. The surface accuracy was computed by means of the relative error Equation (21), where $z_{i,j}$ represents the surface computed via Equation (15), $H_{i,j}$ represents the reference data, and $n \cdot m$ indicates the number of data. In this way, Equation (21) was computed by employing the surface data shown in Figure 7a, and the relative error was 0.691%. From this surface, a flat surface model was generated via multi-objective optimization based on a genetic algorithm by means of Equation (8) and the coordinates $(x_{i,j}, y_{i,j}, z_{i,j})$. The steps of this multi-objective optimization are described as follows.

The first step was to determine the initial population through the coordinates $(x_{i,j}, y_{i,j}, z_{i,j})$. To carry it out, the initial constants (a, b, c) were computed by means of Equation (11) through Equation (13) as described in Section 2.2. Then, four parents $(\mathcal{P}_{1,1}, \mathcal{P}_{2,1}, \mathcal{P}_{3,1}, \mathcal{P}_{4,1})$ were computed by employing the sub-indices $(i = 0, j = 0, \delta = 82, \Delta = 80)$, $(i = 82, j = 0, \delta = 82, \Delta = 80)$, $(i = 0, j = 80, \delta = 82, \Delta = 80)$, and $(i = 82, j = 80, \delta = 82, \Delta = 80)$. In this case, $n = 164$ and $m = 160$. Thus, four constants (a, b, c) were obtained, and they were defined as the first parents $(\mathcal{P}_{1,2}, \mathcal{P}_{2,1}, \mathcal{P}_{3,1}, \mathcal{P}_{4,1})$, where the parent $\mathcal{P}_{1,1}$ is represented by the constants $(a = -0.1061, b = -4.2290, c = 31.1214)$, the parent $\mathcal{P}_{2,1}$ is represented by $(a = 0.5671, b = -3.672, c = 30.8239)$, the parent $\mathcal{P}_{3,1}$ is represented by $(a = 0.8934, b = -1.1235, c = 29.1065)$, and the parent $\mathcal{P}_{4,1}$ is represented by $(a = 0.1432, b = -0.4218, c = 27.8532)$. Then, the terms $(am - 2*as)$ and $(am + 2*as)$ were computed to determine the maximum and minimum of a. Additionally, the terms $(bm - 2*as)$ and $(bm + 2*bs)$ were computed to determine the maximum and minimum of b, and the terms $(cm - 2*cs)$ and $(cm + 2*cs)$ were computed to determine the maximum and minimum of c. Thus, the initial population was completed. Then, the second step was to determine the current children through the crossover by computing Equation (3) through Equation (6) by employing $t = 0$ and $t = 1$. The probability of crossover is deduced via average fitness. Thus, the crossover is performed when the average fitness is enhanced. Next, the third step was to evaluate the fitness by computing Equation (14) using the constants (a, b, c) and the coordinates $(x_{i,j}, y_{i,j}, z_{i,j})$. Then, the fourth step was to select the parents of the next generation from the best current parents and children, where the parents $\mathcal{P}_{1,k+1}$ and $\mathcal{P}_{3,k+1}$ were chosen from the pairs $(\mathcal{P}_{1,k}, \mathrm{P}_{2,k})$ and $(\mathcal{P}_{3,k}, \mathcal{P}_{4,k})$, respectively. In addition, the parents $\mathcal{P}_{2,t+1}$ and $\mathcal{P}_{4,t+1}$ were chosen from the children $(C_{1,k}, C_{2,k}, C_{3,k}, C_{4,k})$ and $(C_{5,k}, C_{6,k}, C_{7,k}, C_{8,k})$, respectively. Then, the fifth step was to mutate the worst parent, which was selected through the fitness Equation (14). Thus, if the new parent enhanced the fitness, the worst parent was mutated. Otherwise, the worst parent was not mutated. In addition, one constant was mutated from a parent, which was selected in random form. To do so, a new constant was randomly generated from the solution space, and it was replaced in the selected parent to compute the fitness via Equation (14). If the new constant improved the fitness, the mutation was carried out; if not, the constant was not mutated. Then, the second step was carried out by computing Equation (3) through Equation (6) to create the children of the $(k + 1)$ generation. Additionally, the fitness of these children was computed via Equation (14). From this step, the population of the $(k + 1)$ generation was completed. The steps to compute the $(k + 1)$ generation were repeated until the multi-objective function in Equation (14) was minimized. In this case, the number of generations to obtain the optimal constants (a, b, c) was 157. The optimal constants were $a = 0.307$, $b = -1.983$, and $c = 28.723$, and they were replaced in Equation (8) to determine the flat surface model $z_{i,j} = 0.307x_{i,j} - 1.983y_{i,j} + 28.723$. This flat surface model produced the flat surface shown in Figure 7c. The fitting accuracy provided by the flat surface model was determined by means of Equation (22) by employing the surface data shown in Figure 7b, where $M_{i,j}$ represents the surface data provided by the flat surface model Equation (8), $z_{i,j}$ represents the surface data recovered via Equation (15), and $n \cdot m$ indicates the data number. Additionally, the optimal gap of the genetic algorithm was determined by computing the relative error Equation (22). From this step, the flat surface model produced a relative error of 1.231% with respect to the paper surface shown in Figure 7b.

The running time is defined by the number of generations. In this case, 157 generations were computed to optimize the micro-scale flat surface model. In this way, the micro-scale flat surface modeling was performed through multi-objective optimization. The efficiency of the multi-objective optimization is described through the parameters in the genetic algorithm as follows. The population size is determined through the constants (a, b, c). For each constant, an initial population of twelve chromosomes was generated. Therefore, one generation included a population of 36 chromosomes. The crossover probability determines how often the crossover is performed. When the average fitness of the parents is improved, the crossover is carried out. In this way, the multi-objective optimization is performed several times on the same flat surface model. The result of the probability of crossover was in the interval from 0.21 to 0.55. The mutation probability determines how often the chromosome can be mutated. Thus, if the new parent improves the fitness, the worst parent is mutated. In the same way, for the parameter mutation, if the new parameter improves the fitness, the parameter is mutated. Thus, the multi-objective optimization is performed several times on the same flat surface model. The result of the probability of mutation was in the interval from 0.26 to 0.53. The number of generations indicates the number of the iterations to obtain the optimal constants. In this case, 157 generations were performed to accomplish the flat surface model. The optimal gap was computed via Equation (22), and the result was a relative error of 1.231%. Based on these results, the genetic algorithm was examined. In this case, the genetic provided good crossover probability and mutation probability. It is because the probability was in the interval established in the surface modeling optimization. The population size is related to the convergence and the number of parameters. In this case, the genetic algorithm provided 12 chromosomes for each parameter. These chromosomes produced results near the optimal solution and reduced iterations. Therefore, the algorithm provided a good population size and number of iterations. The optimal gap established good fitting of the surface model to the target surface.

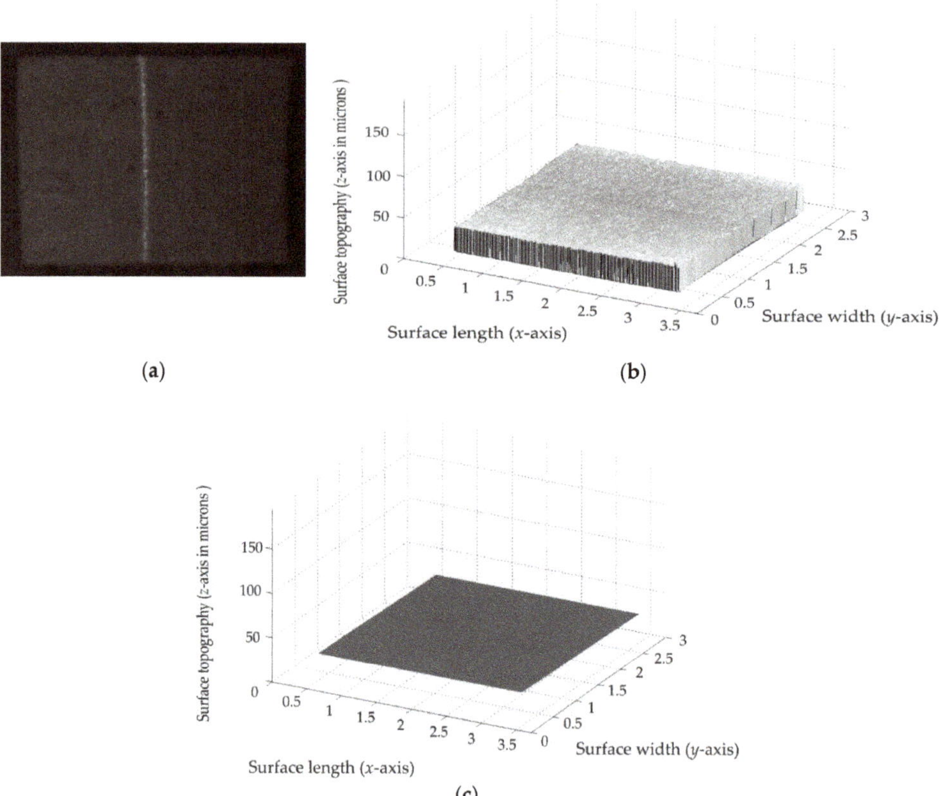

Figure 7. (**a**) Micro laser line projected on flat surface to perform micro-scale flat surface modeling. (**b**) Micro-scale surface retrieved via micro laser line scanning. (**c**) Surface computed by the flat surface model generated via Equation (8).

4. Discussion

The viability of surface modeling in industry 4.0 is determined through the fitting accuracy [42]. In this way, the viability of the micro-scale flat and free-form surface modeling is determined through the fitting accuracy [43,44]. Therefore, the contribution of the proposed flat and free-form surface modeling is deduced from the fitting accuracy with respect to the experimental surface data. Additionally, the algorithm efficiency is included in the contribution. In the proposed multi-objective optimization, the fitting accuracy and efficiency provided good results. The fitting accuracy of the flat and free-form surface modeling was deduced through the quality gap, which was calculated by means of Equation (22). Thus, the micro-scale flat and free-form surface modeling performed via multi-objective optimization fit the surface data with an error smaller than 1.99%. On the other hand, the solution quality and algorithm structure determine algorithm efficiency. The algorithm structure optimized the surface model parameters in good manner. It is because the solution space is determined from the surface data, which provide the solution to accomplish the micro-scale flat and free-form surface models. In this way, the algorithm provided an initial population near the optimal parameters. Therefore, a low error was produced by the algorithm from the first generation. Thus, the algorithm reduced iterations to determine the surface model parameters. This led to providing a running time in a moderated number of iterations. In addition, the multi-objective objective function involves several objective functions to perform the optimization via a genetic algorithm. Thus, several solution candidates were obtained to achieve the convergence through the initial population. Based on these statements, the micro-scale flat and free-form surface modeling via multi-objective optimization enhanced the fitting accuracy of the optical microscope systems. Typically, the microscope systems perform the micro-scale flat and free-form surface modeling with a relative error over 4.6% [45,46], where the flat and free-form surface modeling is optimized through the least-squares method [47,48]. Additionally, the contour data accuracy is related to the fitting accuracy of the surface model. Typically, the optical microscope systems determine the topography data through the gray-level to perform the flat and free-form surface modeling [49,50]. However, the surface profile is not depicted by the gray-level profile. Therefore, the surface modeling via contouring based on gray-level does not provide the best fitting accuracy. Instead, micro laser line reflection depicts the micro-scale surface contour with high accuracy. It is because the laser line reflects the surface contour on the microscope image plane. Moreover, the algorithms of artificial intelligence were implemented in industry 4.0 to perform surface modeling [51]. In the same way, flat and free-form surface modeling has been developed through the algorithms of artificial intelligence [52,53]. These algorithms perform the parameter optimization by employing the traditional search structure [54,55], where the solution space is not defined through the data related to the surface model. Instead, the multi-objective optimization determines the solution space by employing contour data related to the surface model. This procedure provides values that produce a low error from the first generation. Thus, the search procedure begins on a path near the optimal parameters. In this way, the iterations to optimize the model parameters are reduced, and the efficiency is enhanced. This statement is corroborated by the low error produced in the initial population. Thus, the multi-objective optimization performs the optimization by employing contour data. Instead, traditional algorithms generate the initial population in random form [56]. It is because the traditional algorithms do not deduce the search space from the surface data. Furthermore, the multi-objective optimization performs a search inside and outside the parents to avoid the elimination of potential solution candidates. This leads to searching in the entire solution space to find the best solution in efficient form. The viability of the multi-objective optimization via a genetic algorithm is elucidated through the fitting accuracy based on the algorithms of artificial intelligence. To elucidate this, intelligent algorithms were employed to perform flat and free-form surface modeling. Thus, the results provided by these algorithms of artificial intelligence are mentioned as follows. The quality gap provided by ant colony and particle swarm optimization was over 4.23%. These methods performed more than 235 generations to accomplish the flat and free-form surface models. The suitable structure is discussed based on the structure of the particle swarm, which has been implemented to perform surface modeling in industry 4.0. For instance, the particle swarm determines the population through the learning factors, which are described as follows. In this way, the particle swarm performs the population evolution based on two learning factors (α, β), two random numbers (R_1, R_2), and an inertia weight w [57]. Therefore, the particle swarm should compute five variables to determine the population of each generation. Instead, the proposed algorithm computes the variable β through the spread factor α to create the population of each generation. Thus, the suitable structure of the multi-objective optimization via genetic algorithms provides a better efficiency than the particle swarm. In addition, the proposed genetic algorithm computes minus variables than simulated annealing and ant colony to optimize flat and free-form surface models. Based on these statements, the multi-objective optimization based on the genetic algorithm elucidates a suitable structure to construct flat and free-form surface models. This criterion was elucidated through the surface models optimized by particle swarm optimization and genetic algorithms [58]. Thus, the capability of the micro-scale flat and free-form surface modeling via multi-objective optimization is corroborated. Finally, the efficiency of the multi-objective optimization is elucidated through the parameters in the genetic algorithm. The population size is determined by the number of chromosomes of one generation. For instance, the free-form surface model included a population of 6912 chromosomes, and the flat surface model included 36 chromosomes in one generation. However, in both cases, the initial population provided data near the optimal solution. Therefore, a good convergence was achieved. This means that the algorithm provided a good population size in each generation. The crossover probability determines how often the crossover is performed. The mutation probability determines how often the chromosome can be mutated. Based on the results, the genetic algorithm provided a probability of crossover and mutation in and around the interval between 0.25 and 0.6, which was established by recent optimization research. Therefore, the proposed genetic algorithm provided a good crossover probability and mutation probability. The structure of the genetic algorithm reduced iterations. Based on the results described, the genetic algorithm provided a good number of iterations to optimize the model parameters. The optimal gap established good fitting of the surface model to the target surface. It is corroborated with respect to the particle swarm, which performs surface modeling. In addition, the multi-objective optimization via genetic algorithms provides a contribution with respect to our spherical and cylindrical surface modeling, which is referenced in Figure 4. It is because our spherical and cylindrical surface models are represented by just one equation. In this way, the free-form surface model does not fit accurately to the surface when the model is constructed by employing one equation. Therefore, it is necessary to construct the free-form surface model through multi-functions, which fits accurately to surface data. Additionally, the simple optical arrangement provides a low cost to increase the capability of the proposed micro-scale flat and free-form surface modeling. Thus, the multi-objective optimization based on a genetic algorithm and micro laser line projection makes a contribution in the field of flat and free-form surface modeling, which is performed by the optical microscope systems.

The micro-scale flat and free-form surface modeling was implemented in a computer at a velocity of 2.2 GHz. Additionally, the computer performed the control of the CCD camera, which captured 58 images per second. In addition, the slider device was moved by the computer through control software. In this way, the surface contour was recovered in 0.0052 s from each micro laser line image. Additionally, the time to perform the surface modeling was defined by including the surface recovering via laser line scanning. Thus, the free-form surface model of the metallic surface was performed in 67.57 s, and the flat surface model of the paper surface was carried out in 48.36 s.

5. Conclusions

A technique to perform flat and free-form surface modeling through the multi-objective optimization and micro laser line scanning was presented. The multi-objective optimization enhanced the fitting accuracy of the flat and free-form surface models, which was performed by the optical microscope system. This capability was elucidated through the algorithm efficiency and fitting accuracy to optimize flat and free-form surface models. The enhancement of the fitting accuracy was achieved through the algorithm structure, which accurately approximated the model to the micro-scale flat and free-form surfaces. Additionally, the improvement in fitting accuracy was given by the micro laser line scanning. It is because the laser line accurately reflects the surface topography on the microscope image plane. In this way, the micro-scale flat and free-form surface models were constructed through the real surface contour. The multi-objective efficiency was achieved through the solution space, which provided initial solutions with low error to achieve the optimal surface model. Thus, the iterations were reduced to accurately fit the flat and free-form surface models. Thus, the multi-objective optimization via a genetic algorithm proves valuable in industry 4.0 to perform flat and free-form surface modeling in the field of optical microscope systems. In addition, the optical microscope arrangement included simple components such as: an optical microscope, a diode laser, and a CCD camera. This led to reduced cost of the optical arrangement, and it corroborates the capability of the micro-scale flat and free-form surface modeling. Thus, the multi-objective optimization via genetic algorithms and micro laser line scanning was implemented to perform micro-scale flat and free-form surface modeling, with good results in the micro-scale interval.

Funding: This research received no external funding.

Conflicts of Interest: The authors declare no conflict of interest.

References

1. Ghosh, A.K.; Ullah, A.S.; Kubo, A.; Akamatsu, T.; D'Addona, D.M. Machining phenomenon twin construction for Industry 4.0: A Case of Surface Roughness. *J. Manuf. Mater. Process.* **2020**, *4*, 11. [CrossRef]
2. Villalonga, A.; Beruvides, G.; Castaño, F.; Haber, R.E. Cloud-Based Industrial Cyber–physical system for data-driven reasoning: A review and use case on an Industry 4.0 Pilot Line. *IEEE Trans. Ind. Inform.* **2020**, *16*, 5975–5984. [CrossRef]
3. Gan, B.; Zhang, C.; Chen, Y.; Chen, Y.C. Research on role modeling and behavior control of virtual reality animation interactive system in Internet of Things. *J. Real-Time Image Process.* **2021**, *18*, 1069–1083. [CrossRef]
4. Nasr, M.M.; Anwar, S.; Al-Samhan, A.M.; Ghaleb, M.; Dabwan, A. Milling of graphene reinforced Ti6Al4V nanocomposites: An artificial intelligence based industry 4.0 approach. *Materials* **2020**, *13*, 5707. [CrossRef] [PubMed]
5. Majstorovic, V.; Stojadinovic, S.; Zivkovic, S.; Djurdjanovic, D.; Jakovljevic, Z.; Gligorijevic, N. Cyber-physical manufacturing metrology model (CPM3) for sculptured surfaces–turbine blade application. *Procedia CIRP* **2017**, *63*, 658–663. [CrossRef]
6. Elhoone, H.; Zhang, T.; Anwar, M.; Desai, S. Cyber-based design for additive manufacturing using artificial neural networks for Industry 4.0. *Int. J. Prod. Res.* **2020**, *58*, 2841–2861. [CrossRef]
7. Czimmermann, T.; Chiurazzi, M.; Milazzo, M.; Roccella, S.; Barbieri, M.; Dario, P. An autonomous robotic platform for manipulation and inspection of metallic Surfaces in Industry 4.0. *IEEE Trans. Autom. Sci. Eng.* **2022**, *19*, 1691–1706. [CrossRef]
8. Mishra, D.; Roy, R.B.; Dutta, S.; Pal, S.K.; Chakravarty, D. A review on sensor based monitoring and control of friction stir welding process and a roadmap to Industry 4.0. *J. Manuf. Process.* **2018**, *36*, 373–397. [CrossRef]
9. Yan, J.; Meng, Y.; Lu, L.; Li, L. Industrial big data in an Industry 4.0 environment: Challenges, schemes, and applications for predictive maintenance. *IEEE Access* **2017**, *5*, 23484–23491. [CrossRef]
10. Nadolny, K.; Kapłonek, W. Analysis of flatness deviations for austenitic stainless steel work pieces after efficient surface machining. *Meas. Sci. Rev.* **2014**, *14*, 204–212. [CrossRef]
11. Fei, W.; Narsilio, G.A.; Disfani, M.M. Impact of three-dimensional sphericity and roundness on heat transfer in granular materials. *Powder Technol.* **2019**, *355*, 770–781. [CrossRef]
12. Zheng, Y. A simple unified branch-and-bound algorithm for minimum zone circularity and sphericity errors. *Meas. Sci. Technol.* **2020**, *31*, 045005. [CrossRef]
13. Lei, L.; Xie, Z.; Zhu, H.; Guan, Y.; Kong, M.; Zhang, B.; Fu, Y. Improved particle swarm optimization algorithm based on a three–dimensional convex hull for fitting a screw thread central axis. *IEEE Access* **2021**, *9*, 4902–4910. [CrossRef]
14. Brad, E.; Brad, S. Algorithm for designing reconfigurable equipment to enable Industry 4.0 and circular economy-driven manufacturing systems. *Appl. Sci.* **2021**, *11*, 4446. [CrossRef]
15. Ding, Y.; Zhang, W.; Yu, L.; Lu, K. The accuracy and efficiency of GA and PSO optimization schemes on estimating reaction kinetic parameters of biomass pyrolysis. *Energy* **2019**, *176*, 582–588. [CrossRef]
16. Lodewijks, G.; Cao, Y.; Zhao, N.; Zhang, H. Reducing CO_2 emissions of an airport baggage handling transport system using a particle swarm optimization algorithm. *IEEE Access* **2021**, *9*, 121894–121905. [CrossRef]
17. Saini, D.; Kumar, S.; Singh, M.K.; Ali, M. Two view NURBS reconstruction based on GACO model. *Complex Intell. Syst.* **2021**, *7*, 2329–2346. [CrossRef]
18. Tran, N.-H.; Park, H.S.; Nguyen, Q.V.; Hoang, T.D. Development of a smart cyber-physical manufacturing system in the Industry 4.0 Context. *Appl. Sci.* **2019**, *9*, 3325. [CrossRef]
19. Soori, M.; Asmael, M. A review of the recent development in machining parameter optimization. *Jordan J. Mech. Ind. Eng.* **2022**, *16*, 205–223.
20. Ye, X.; Luo, L.; Hou, L.; Duan, Y.; Wu, Y. Laser ablation manipulator coverage path planning method based on an improved ant colony algorithm. *Appl. Sci.* **2020**, *10*, 8641. [CrossRef]
21. Pathak, V.K.; Nayak, C.; Singh, R.; Dikshit, M.K.; Sai, T. Optimizing parameters in surface reconstruction of transtibial prosthetic socket using central composite design coupled with fuzzy logic-based model. *Neural Comput. Appl.* **2020**, *32*, 15597–15613. [CrossRef]
22. Kannan, K.; Arunachalam, N.; Chawla, A.; Natarajan, S. Multi-sensor data analytics for grinding wheel redress life estimation- an approach towards Industry 4.0. *Procedia Manuf.* **2018**, *26*, 1230–1241. [CrossRef]
23. Florková, Z.; Jambor, M. Quantification of aggregate surface texture based on three dimensional microscope measurement. *Procedia Eng.* **2017**, *192*, 195–200. [CrossRef]
24. Bustillo, A.; Pimenov Yu, D.; Mia, M.; Kapłonek, W. Machine-learning for automatic prediction of flatness deviation considering the wear of the face mill teeth. *J. Intell. Manuf.* **2021**, *32*, 895–912. [CrossRef]
25. Melentiev, R.; Fang, F. Tailoring of surface topography for tribological purposes by controlled solid particle impacts. *Wear* **2020**, *444–445*, 203164. [CrossRef]

26. Haitjema, H. Straightness, flatness and cylindricity characterization using discrete Legendre polynomials. *CIRP Ann.-Manuf. Technol.* **2020**, *69*, 457–460. [CrossRef]
27. Vidal, M.; Ostra, M.; Imaz, N. García-Lecina, E.; Ubide, C. Feature descriptors from scanned images of chromium electrodeposits as predictor parameters of surface roughness and crystallographic texture. *Chemom. Intell. Lab. Syst.* **2015**, *149*, 90–98. [CrossRef]
28. Shaukat, N.; Ahmad, A.; Mohsin, B.; Khan, R.; Ud-Din Khan, S.; Ud-Din Khan, S. Multi objective core reloading pattern optimization of PARR-1 using modified genetic algorithm coupled with Monte Carlo methods. *Sci. Technol. Nucl. Install.* **2021**, *2021*, 1802492. [CrossRef]
29. Melin, P.; Sánchez, D. Multi-objective optimization for modular granular neural networks applied to pattern recognition. *Inf. Sci.* **2018**, *460–461*, 594–610. [CrossRef]
30. Akleman, E.; Srinivasan, V.; Chen, J. Interactive modeling of smooth manifold meshes with arbitrary topology: G^1 stitched bi-cubic Bézier patches. *Comput. Graph.* **2017**, *66*, 64–73. [CrossRef]
31. Long, Q.; Wu, C.; Wang, X.; Huang, T.; Wang, X. A genetic algorithm for unconstrained multi-objective optimization. *Swarm Evol. Comput.* **2015**, *22*, 1–14. [CrossRef]
32. Jia, L.; Cheng, D.; Chiu, M.S. Pareto-optimal solutions based multi-objective particle swarm optimization control for batch processes. *Neural Comput. Appl.* **2012**, *21*, 1107–1116. [CrossRef]
33. Hazra, J.; Sinha, A.K. Congestion Management Using Multiobjective Particle Swarm Optimization. *IEEE Trans. Power Syst.* **2007**, *22*, 1726–1734. [CrossRef]
34. Tan, K.C.; Chiam, S.C.; Mamun, A.A.; Goh, C.K. Balancing exploration and exploitation with adaptive variation for evolutionary multi–objective optimization. *Eur. J. Oper. Res.* **2009**, *197*, 701–713. [CrossRef]
35. Li, C.Y.; Zhu, C.G. G^1 continuity of four pieces of developable surfaces with Bézier boundaries. *J. Comput. Appl. Math.* **2018**, *329*, 164–172. [CrossRef]
36. Tao, P.; Sun, Z.; Sun, Z. An improved intrusion detection algorithm based on GA and SVM. *IEEE Access* **2018**, *6*, 13624–13631. [CrossRef]
37. Muñoz-Rodríguez, J.A. Micro-scale spherical and cylindrical surface modeling via metaheuristic algorithms and micro laser line projection. *Algorithms* **2022**, *15*, 145. [CrossRef]
38. Muñoz-Rodríguez, J.A.; Rodríguez-Vera, R. Evaluation of the light line displacement location for object shape detection. *J. Mod. Opt.* **2003**, *50*, 137–154. [CrossRef]
39. Yu, H.; Huang, Q.; Zhao, J. Fabrication of an optical fiber micro-sphere with a diameter of several tens of micrometers. *Materials* **2014**, *7*, 4878–4895. [CrossRef]
40. Lin, T.W.; Wang, C.H. A hybrid genetic algorithm to minimize the periodic preventive maintenance cost in a series-parallel system. *J. Intell. Manuf.* **2012**, *23*, 1225–1236. [CrossRef]
41. Gibbs, M.S.; Maier, H.R.; Dandy, G.C. Relationship between problem characteristics and the optimal number of genetic algorithm generations. *Eng. Optim.* **2011**, *43*, 349–376. [CrossRef]
42. Cheng, X. Algorithm of CAD surface generation for complex pipe model in Industry 4.0 background. *Comput. Intell. Neurosci.* **2022**, *2022*, 7062052. [CrossRef] [PubMed]
43. Kovács, Z.F.; Viharos, Z.J.; Kodácsy, J. Surface flatness and roughness evolution after magnetic assisted ball burnishing of magnetizable and non-magnetizable materials. *Measurement* **2020**, *158*, 107750. [CrossRef]
44. Rena, M.J.; Suna, L.J.; Liub, M.Y.; Cheungb, C.F.; Yina, Y.H.; Cao, Y.L. A weighted least square based data fusion method for precision measurement of freeform surfaces. *Precis. Eng.* **2017**, *48*, 144–151. [CrossRef]
45. Miko, B. Assessment of flatness error by regression analysis. *Measurement* **2021**, *171*, 108720. [CrossRef]
46. Mikó, B.; Varga, B.; Zębala, W. The effect of the feed direction on the micro- and macro accuracy of 3D ball-end milling of chromium-molybdenum Alloy Steel. *Materials* **2019**, *12*, 4038. [CrossRef]
47. Ye, L.; Qian, J.; Haitjema, H.; Reynaerts, D. On-machine chromatic confocal measurement for micro-EDM drilling and milling. *Precis. Eng.* **2022**, *76*, 110–123. [CrossRef]
48. Zhu, D.; Feng, X.; Xu, X.; Yang, Z.; Li, W.; Yan, S.; Ding, H. Robotic grinding of complex components: A step towards efficient and intelligent machining–challenges, solutions, and applications. *Robot. Comput.-Integr. Manuf.* **2020**, *65*, 101908. [CrossRef]
49. Fang, H.; Zhu, H.; Yang, J.; Huang, X.; Hu, X. Three-dimensional angularity characterization of coarse aggregates based on experimental comparison of selected methods. *Part. Sci. Technol.* **2021**, *1*, 1–9. [CrossRef]
50. Chen, L.C. Automatic 3D surface reconstruction and sphericity measurement of micro spherical balls of miniaturized coordinate measuring probes. *Meas. Sci. Technol.* **2007**, *18*, 1748–1755. [CrossRef]
51. Stojadinovic, S.M.; Majstorovic, V.D.; Gąska, A.; Sładek, J.; Durakbasa, N.M. Development of a coordinate measuring machine—based inspection Planning system for industry 4.0. *Appl. Sci.* **2021**, *11*, 8411. [CrossRef]
52. Shu, H.; Zou, C.; Chen, J.; Wang, S. Research on micro/nano surface flatness evaluation method based on improved particle swarm optimization algorithm. *Front. Bioeng. Biotechnol.* **2021**, *9*, 775455. [CrossRef]
53. Yang, Y.; Dong, Z.; Meng, Y.; Shao, C. Data-driven intelligent 3D surface measurement in smart manufacturing: Review and outlook. *Machine* **2021**, *9*, 13. [CrossRef]
54. Liang, Z.; Liu, X.; Ye, B.; Wang, Y. Performance investigation of fitting algorithms in surface micro-topography grinding processes based on multi-dimensional fuzzy relation set. *Int. J. Adv. Manuf. Technol.* **2013**, *67*, 2779–2798. [CrossRef]

55. Tian, X.; Kong, L.; Kong, D.; Yuan, L.; Kong, D. An improved method for NURBS surface based on particle swarm optimization BP neural network. *IEEE Access* **2020**, *8*, 184656–184663. [CrossRef]

56. Wu, Z.; Wang, X.; Fu, Y.; Shen, J.; Jiang, Q.; Zhu, Y.; Zhou, M. Fitting scattered data points with ball B-Spline curves using particle swarm optimization. *Comput. Graph.* **2018**, *72*, 1–11. [CrossRef]

57. Tsiptsis, I.N.; Liimatainen, L.; Kotnik, T.; Niiranen, J. Structural optimization employing isogeometric tools in Particle Swarm Optimizer. *J. Build. Eng.* **2019**, *24*, 100761. [CrossRef]

58. Wang, S.; Xia, Y.; Wang, R.; You, L.; Zhang, J. Optimal NURBS conversion of PDE surface-represented high-speed train heads. *Optim. Eng.* **2019**, *20*, 907–928. [CrossRef]

MDPI

St. Alban-Anlage 66

4052 Basel

Switzerland

Tel. +41 61 683 77 34

Fax +41 61 302 89 18

www.mdpi.com

Energies Editorial Office

E-mail: energies@mdpi.com

www.mdpi.com/journal/energies

9 783036 554877